全国高等医药院校药学类实验教材

无机化学实验

（第二版）

主　编　刘迎春
主　审　王国清
副主编　赵　兵　刘晶莹
编　者　（以姓氏笔画为序）
　　　　丁怀伟　王绍宁　王鸿钢
　　　　兰　阳　刘迎春　刘晶莹
　　　　张　莹　赵　兵　段丽颖
　　　　凌俊红

中国医药科技出版社

内 容 提 要

本书为全国高等医药院校药学类实验教材之一。全书分为6章，分别为基本知识和基本操作、基本操作训练实验、化学原理实验、元素化学实验、无机化合物制备实验和前沿性实验和综合设计实验。为适应教育国际化的要求，从第二章开始增加了英文对照内容，以便于学生在阅读英文文献、撰写英文论文时参考。

本书可供高等医药院校药学类及相关专业使用，也可作为医药行业相关人员培训用书。

图书在版编目（CIP）数据

无机化学实验/刘迎春主编．—2 版．—北京：中国医药科技出版社，2014.8
全国高等医药院校药学类实验教材
ISBN 978－7－5067－6908－2

Ⅰ．①无… Ⅱ．①刘… Ⅲ．①无机化学－化学实验－医学院校－教材 Ⅳ．①O61－33

中国版本图书馆 CIP 数据核字（2014）第 160574 号

美术编辑 陈君杞
版式设计 郭小平

出版 中国医药科技出版社
地址 北京市海淀区文慧园北路甲 22 号
邮编 100082
电话 发行：010－62227427 邮购：010－62236938
网址 www.cmstp.com
规格 787×1092mm $\frac{1}{16}$
印张 14
字数 280 千字
初版 2006 年 3 月第 1 版
版次 2014 年 8 月第 2 版
印次 2022 年 8 月第 7 次印刷
印刷 三河市航远印刷有限公司
经销 全国各地新华书店
书号 ISBN 978－7－5067－6908－2
定价 32.00 元

本社图书如存在印装质量问题请与本社联系调换

全国高等医药院校药学类规划教材常务编委会

名誉主任委员　邵明立　林蕙青
主　任　委　员　吴晓明（中国药科大学）
副主任委员　（按姓氏笔画排序）
　　　　　　　刘俊义（北京大学药学院）
　　　　　　　匡海学（黑龙江中医药大学）
　　　　　　　朱依谆（复旦大学药学院）
　　　　　　　朱家勇（广东药学院）
　　　　　　　毕开顺（沈阳药科大学）
　　　　　　　吴少祯（中国医药科技出版社）
　　　　　　　吴春福（沈阳药科大学）
　　　　　　　张志荣（四川大学华西药学院）
　　　　　　　姚文兵（中国药科大学）
　　　　　　　高思华（北京中医药大学）
　　　　　　　彭　成（成都中医药大学）
委　　　员　（按姓氏笔画排序）
　　　　　　　王应泉（中国医药科技出版社）
　　　　　　　田景振（山东中医药大学）
　　　　　　　李　高（华中科技大学同济药学院）
　　　　　　　李元建（中南大学药学院）
　　　　　　　李青山（山西医科大学药学院）
　　　　　　　杨　波（浙江大学药学院）
　　　　　　　杨世民（西安交通大学药学院）
　　　　　　　陈思东（广东药学院）
　　　　　　　侯爱君（复旦大学药学院）
　　　　　　　娄红祥（山东大学）
　　　　　　　宫　平（沈阳药科大学）
　　　　　　　祝晨蔯（广州中医药大学）
　　　　　　　柴逸峰（第二军医大学药学院）
　　　　　　　黄　园（四川大学华西药学院）
　　　　　　　朱卫丰（江西中医药大学）
秘　　　书　夏焕章（沈阳药科大学）
　　　　　　　徐晓媛（中国药科大学）
　　　　　　　沈志滨（广东药学院）
　　　　　　　浩云涛（中国医药科技出版社）
　　　　　　　赵燕宜（中国医药科技出版社）

第二版前言

本书可用作高等药学院校各专业或普通高等院校化学专业的无机化学实验教材。

药学和化学就其本质而言均为实验科学，无机化学又是学生进入大学接触的首门专业基础课程，无机化学实验教学可以帮助学生加深理解无机化学基本原理及元素主要化学性质，熟悉无机化合物的常用制备手段和分析方法，训练实验操作的基本技能和技巧，培养科学思维，增强创新意识和创新能力，对提高学生综合素质具有重要意义。

作为我国历史悠久的综合性药学院校，沈阳药科大学始终以坚持为国家培养符合社会发展需求的高端药学人才为宗旨，不断努力构建内容先进、特色鲜明的理论和实验教学课程体系。

本实验教材第一版于2004年由中国医药科技出版社出版，在多年教学应用中，得到广大师生的充分肯定，同时编者也发现了许多不足之处，因此，在本次编写中将实验精炼为27个，包括基本操作训练实验、化学原理实验、元素化学实验、无机化合物的制备实验、前沿性实验和综合设计实验五大部分。

本实验教材尽力突出以下特点：

1. 增加药学特色实验，注意与我国药学发展现状的实际相结合，展现无机化学在药学领域中的应用。在保留"药用氯化钠的制备及杂质限度检查"实验的基础上，增加了两个无机药物的合成及检查（药用氢氧化铝、药用碱式碳酸铋），不仅可以激发学生对药学专业的浓厚兴趣，也为后续的药物合成及药物分析课程的学习奠定基础。

2. 配合无机化学理论教学，内容覆盖无机化学的各个主要方面，包括化学热力学、化学动力学、四大化学平衡、原子结构与分子结构以及元素化学。

3. 重视综合设计实验以外，还增加了前沿性实验——两个无机纳米粒子的合成实验（纳米二氧化钛、纳米四氧化三铁），旨在培养学生综合运用前期理论和实验知识，独立分析问题和解决问题，培养团队合作精神，感受前沿科技的发展，激发科学研究的热情。

4. 编写双语体系，以适应21世纪教育、科技和社会发展的需要，帮助学生在本科教育阶段熟练掌握专业英语，特别是实验技能方面实用英语，为将来的继续深造和国际学术交流打下基础。

本教材由王国清教授主审，刘迎春、赵兵、刘晶莹、王绍宁、张莹、王鸿钢、段丽颖、兰阳、凌俊红及丁怀伟老师参加编写，夏丹丹、肖琰两位年轻博士对本书英文部分进行核审及修改,本教材是无机化学教研室全体同仁辛勤劳动、努力创新的结晶。另外,本校 2011 级应用化学专业的吴洋、赵小强两位同学为本书的插图做了大量工作,在此一并表示感谢!

由于时间紧迫和编者水平有限,教材中的错误和不妥之处在所难免,欢迎兄弟院校师生批评指正,以期完善和提高。

编者

2014 年 7 月

目 录

绪论 ……………………………………………………………………………………… (1)
第一章　基本知识和基本操作 …………………………………………………………… (5)
第二章　基本操作训练实验 ……………………………………………………………… (33)
Chapter 2　Basic Operation Training Experiments …………………………………… (33)
　实验一　称量练习 …………………………………………………………………… (33)
　Experiment 1　Weighing Exercise ………………………………………………… (34)
　实验二　容量仪器的校正 …………………………………………………………… (37)
　Experiment 2　Calibration of Volumetric Glassware …………………………… (38)
　实验三　酸碱滴定练习 ……………………………………………………………… (41)
　Experiment 3　Acid-Base Titration Exercise …………………………………… (42)
　实验四　硝酸钾溶解度曲线的绘制 ………………………………………………… (45)
　Experiment 4　Drawing of the Solubility Curve of Potassium Nitrate ………… (48)
第三章　化学原理实验 …………………………………………………………………… (52)
Chapter 3　Chemical Principle Experiments …………………………………………… (52)
　实验五　氯化铵摩尔生成焓的测定 ………………………………………………… (52)
　Experiment 5　Determining the Molar Formation Enthalpy of Ammonium Chloride
　　　　　　　　……………………………………………………………………… (55)
　实验六　化学反应速率和活化能的测定 …………………………………………… (58)
　Experiment 6　Determining the Rate and Activation Energy of Chemical Reaction
　　　　　　　　……………………………………………………………………… (61)
　实验七　凝固点降低法测定葡萄糖的摩尔质量 …………………………………… (65)
　Experiment 7　Determination of the Molar Mass of Glucose by Freezing Point Depression
　　　　　　　　Method ……………………………………………………………… (68)
　实验八　电位法测定硼酸的电离常数 ……………………………………………… (71)
　Experiment 8　Determination of Dissociation Constant of a Weak Acid by
　　　　　　　　Potentiometric Method …………………………………………… (73)
　实验九　电动势法测定氯化银的溶度积常数 ……………………………………… (75)
　Experiment 9　Determining Solubility Product Constant of Silver Chloride by Cell
　　　　　　　　Potential Method …………………………………………………… (76)

· 1 ·

实验十　酸碱平衡和沉淀溶解平衡 ……………………………………………… (78)

Experiment 10　Acid-Base Equilibrium and Precipitation-Dissolution Equilibrium

…………………………………………………………………………………… (82)

实验十一　氧化还原反应 ………………………………………………………… (86)

Experiment 11　Redox Reactions …………………………………………………… (90)

实验十二　分光光度法测定磺基水杨酸合铁（Ⅲ）配合物的组成和稳定常数

…………………………………………………………………………………… (95)

Experiment 12　Determining the Composition and Stability Constant for Iron（Ⅲ）-

Sulfosalicylic Acid Coordination Compound by Spectrophotometry

…………………………………………………………………………………… (97)

实验十三　配合物的生成和性质 ………………………………………………… (101)

Experiment 13　Formation and Properties of Coordination Compounds ………… (104)

实验十四　某些无机物分子或基团的空间构型 ………………………………… (108)

Experiment 14　Steric Configuration of Certain Inorganic Molecules or Groups …… (110)

第四章　元素化学实验 …………………………………………………………… (114)

Chapter 4　Elements Chemistry Experiments ………………………………… (114)

实验十五　p 区元素 ……………………………………………………………… (114)

Experiment 15　p Block Elements ………………………………………………… (117)

实验十六　d 区元素和 ds 区元素 ………………………………………………… (128)

Experiment 16　d Block Elements and ds Block Elements ……………………… (131)

实验十七　常见无机离子的鉴定反应 …………………………………………… (139)

Experiment 17　Identification Reactions of the Common Inorganic Ions ………… (144)

实验十八　样品分析 ……………………………………………………………… (151)

Experiment 18　Analysis of Samples ……………………………………………… (155)

第五章　无机化合物制备实验 …………………………………………………… (161)

Chapter 5　Preparation Experiments of Inorganic Compounds ……………… (161)

实验十九　药用氯化钠的制备及杂质限度检查 ………………………………… (161)

Experiment 19　Preparation of Medicinal Sodium Chloride and Examination of the

Limits of Impurities …………………………………………………… (165)

实验二十　硫酸亚铁铵的制备 …………………………………………………… (170)

Experiment 20　Preparation of Ferrous Ammonium Sulphate …………………… (172)

实验二十一　药用氢氧化铝的制备、鉴别、制酸力检查及含量测定 ………… (175)

Experiment 21　Preparation, Identification, Neutralizing Capacity Examination

and Content Assay of Medicinal Aluminum Hydroxide …………… (176)

实验二十二　药用碱式碳酸铋的制备、鉴别、制酸力检查及含量测定 ……… (178)

Experiment 22 Preparation, Identification, Neutralizing Capacity Examination and Content Assay of Medicinal Bismuth Subcarbonate ……………… (180)

第六章　前沿性实验和综合设计实验 ……………………………………… (183)
Chapter 6 Frontier Experiments and Comprehensive Designing Experiments
……………………………………………………………………… (183)

  实验二十三　二氧化钛纳米粒子的制备及其光催化活性的测定 ……… (183)
  Experiment 23 Preparation of Nano TiO_2 Particles and Determination of Its Photocatalytic Activity ……………………………………… (184)

  实验二十四　纳米四氧化三铁的化学共沉淀法制备及表征 …………… (186)
  Experiment 24 Preparation and Characterization of Nano Fe_3O_4 by Chemical Coprecipitation ……………………………………………… (188)

  实验二十五　十二钨硅酸的制备、萃取分离及表征 …………………… (189)
  Experiment 25 The preparation, extraction and characterization of dodecatungstosilicic acid ……………………………………………………………… (191)

  实验二十六　三氯化六氨合钴（Ⅲ）的制备及组成的测定 …………… (193)
  Experiment 26 Preparation and Component Analysis of $[Co(NH_3)_6]Cl_3$ ………… (196)

  实验二十七　植物中某些元素的分离与鉴定 …………………………… (199)
  Experiment 27 Isolation and Identification of Inorganic Elements in Plants ……… (202)

附录 …………………………………………………………………………… (205)
  一、常用的酸碱指示剂 …………………………………………………… (205)
  二、常用的酸碱密度和浓度 ……………………………………………… (205)
  三、酸度计简介 …………………………………………………………… (205)
  四、分光光度计简介 ……………………………………………………… (208)
  五、电导率仪 ……………………………………………………………… (210)

绪　　论

化学是一门研究物质的组成、结构、性质以及变化规律的基础自然科学，是以实验为基础的重要学科，通过化学实验能够获得生动的感性知识，从而能够更好地理解和巩固所学的化学知识，特别是在培养具有创新意识和创新能力的素质教育中，实验更加突出地占有相当重要的地位。实验是化学的灵魂，是化学魅力之所在，化学实验是培养和发展思维能力及创新能力的重要途径，同时化学实验也是检验化学理论正确与否的唯一标准。中国核弹先驱，三次与诺贝尔奖擦肩而过的著名物理学家王淦昌先生说过，敢于大胆设想是第一位的，只有这样，才能创出新路，但光有新思路还不够，最重要的是要自己动手做实验，验证自己的想法。因此，学习化学必须重视实验的重大作用，在实验中不断体会、理解和创造的过程，不断形成化学科学创新的意识和严谨的科学态度，并努力去尝试创新，创造出更多的生产力。

一、无机化学实验目的

无机化学实验是药学类院校开设的第一门实验课，与无机化学理论课有紧密的联系。其主要目的是：

（1）学生经过基本实验的严格训练，能够规范地掌握实验的基本操作、基本技术和基本技能。

（2）学生通过做实验，可以直接获得大量物质变化的感性认识，经归纳、总结，从感性认识上升到理性认识，从而加深对理论课中基本原理和基本知识的理解。

（3）学生在微型化的仪器装置中进行的微型实验，可以用尽可能少的试剂来获取尽可能多的化学信息。培养学生环境保护意识和实验室安全意识。

（4）在设计合成实验及基本原理实验教学中，增加了仪器实验的内容。通过建立研究对象和测试方法的联系，开阔学生的视野，为学生未来发展做好科学储备。

（5）在基本实验训练的基础上，开设综合设计实验，要求学生自己提出问题、查阅资料、设计方案、实验操作、记录实验现象、分析实验结果并讨论。通过化学实验的全过程，使学生得到全面有效的训练，逐步具备分析问题、解决问题的能力。

（6）在培养智力因素的同时，化学实验又是培养学生科学素养的理想场所。实验环节，不仅有利于培养学生整洁、节约、有条不紊的实验素养，而且可以训练学生勤奋好学、乐于协作、实事求是的科学品德和科学精神。

二、无机化学实验学习方法

学生在教师指导下独立完成无机化学实验。为获得较好的实验效果，达到预期实验目的，需要有正确的学习态度和学习方法。具体应做到以下几点：

1. 预习

（1）应认真阅读实验教材，明确实验目的和实验原理，熟悉实验内容、主要操作步骤及数据处理方法，并提出注意事项，合理安排时间。对实验中涉及的基本操作及有关仪器的使用，也要做到预习。

（2）应根据实验内容查阅附录及有关资料，记录实验所需的物理化学数据、定量实验的计算公式及反应方程式等，认真写好预习报告。注意在报告中预留记录实验现象和数据的位置。对于没有达到上述预习要求者，不准参加本次实验。

2. 实验

（1）按教材规定的实验内容规范操作，仔细观察实验现象，认真测定数据，将数据如实记录在预习报告中，不得随意更改、删减。这是培养良好科学习惯的重要环节。

（2）实验中要勤于思考、细心观察，自己分析、解决问题。对实验现象有疑惑，或实验结果误差太大，要认真分析操作过程，努力找到原因。如果必要，可以在教师指导下，做对照实验、空白实验，或自行设计实验进行核实。以培养分析问题、解决问题的能力。

（3）如实验失败，要查明原因，经教师准许后重做实验。

3. 实验报告　实验报告是对本次实验的概括和总结，是对实验记录进行整理，对相关理论知识加深理解的过程。

（1）实验现象要表述正确，并进行合理的解释，写出相应的反应式，得出结论。

（2）对实验数据进行处理（包括计算、作图、误差的表示等）。

（3）分析产生误差的原因。针对实验中遇到的疑难问题提出自己的见解，包括对实验方法、教学方法和实验内容提出改进意见或建议。

（4）实验报告要按一定的格式书写，字迹端正，表格清晰，图形规范，叙述要简明扼要。这是培养严谨的科学态度和实事求是科学精神的重要措施。

三、无机化学实验室规则

严格遵守实验室规则，有利于形成整洁、节约、有条不紊等良好的实验素养。具体内容如下：

（1）实验前认真预习，明确目的要求，了解实验的步骤、方法和基本原理。

（2）实验时按学号对号入座，严格遵守操作规则，保证整个实验过程安全。

（3）实验过程中必须保持肃静，不准讨论与实验无关的内容。不迟到、不早退。

（4）爱护仪器设备，注意节约水电。若破损仪器，立即报告指导教师，填写破损单后到库房领取。

（5）公用仪器（药品）在原处使用，不得挪动。

（6）使用精密仪器时，必须严格按照操作规程进行操作，细心谨慎，以免损坏仪器。如发现仪器有故障，应立即停止使用，报告指导教师，及时排除故障。

（7）使用药品时，应按规定量取用，如果书中未规定用量，应注意节约，尽量少用。液体药品一经取出，就不能倒回原试剂中，以免污染药品。自备滴管只限于在试管或烧杯中转移药品，不能从公用试剂瓶中取药，以免污染公用药品。

(8) 实验过程中要随时保持清洁，用过的火柴梗、废纸片要丢入废物盘内，不能丢入水槽，以免堵塞。

(9) 实验结束，必须把实验台、仪器设备整理好，药品摆放整齐，关闭水电。经教师检查后，方可离开实验室。

四、无机化学实验室安全操作

化学实验过程中，学生经常要用到水、电和各种化学药品。由于化学药品多是易燃、易爆和有腐蚀性的，因此实验室潜藏着各种事故发生的隐患。因而，重视安全操作，掌握一般性救护措施是非常必要的。

1. 实验室安全规则

(1) 水、电和酒精灯使用完毕，要立即关闭。

(2) 浓酸、浓碱、洗液、液溴及其他有具有强腐蚀性的液体，不要洒在皮肤和衣服上。稀释硫酸时，必须将酸倒入水中，切勿将水注入硫酸中，以免迸溅。

(3) 能产生有毒或有刺激性气体的实验，要在通风橱内进行。如加热盐酸和硝酸，或使用强酸和强碱溶解、分解试样的时候，均应该在通风橱内进行。

(4) 有毒试剂，如：重铬酸钾、钡盐、铅盐、砷的化合物、汞及汞的化合物，特别是氰化物，不得进入口内或接触伤口。剩余的废液也不能随意倒入下水道。用剩的有毒药品应交还指导教师。

(5) 金属汞易挥发，吸入体内逐渐累积将引起慢性中毒。使用时要特别小心。一旦洒落，要尽可能收集起来，并用硫粉覆盖在洒落处，使之转化为硫化汞。

(6) 钠、钾、白磷等暴露在空气中易燃烧。故钠、钾保存在煤油中，白磷保存在水中，取用时用镊子夹取。

(7) 绝对不允许随意混合各种化学药品，以免发生意外事故。氯酸钾、高锰酸钾等强氧化剂或其混合物不能研磨，以免爆炸。

(8) 注意保护眼睛，必要时带防护镜。防止眼睛受刺激性气体的熏染，更要防止化学药品等异物进入眼内。

(9) 严禁在实验室内饮食、吸烟。严禁在实验室穿拖鞋。实验完毕，应洗净双手，再离开实验室。

2. 化学实验意外事故处理

(1) 烫伤　烫伤后切勿用冷水冲洗。如伤处皮肤未破，在伤口处抹烫伤油膏或万花油。如伤处皮肤已破，可涂 10% $KMnO_4$ 溶液润湿伤口再抹烫伤膏。

(2) 割伤　应先挑出伤口中的异物。轻伤可在伤口上涂紫药水，再用消毒纱布包扎。伤口较重，应立即到医院医治。

(3) 受酸腐蚀　先用大量水冲洗，再用饱和碳酸氢钠或稀氨水冲洗，最后再用水冲洗。如溅入眼中，立即先用大量水冲洗，再用 1% 碳酸氢钠溶液冲洗。

(4) 受碱腐蚀　先用大量水冲洗，再用醋酸溶液（20g/L）或硼酸溶液冲洗，最后再用水冲洗。如溅入眼中，可先用硼酸溶液冲洗，再用大量水冲洗。

(5) 受溴灼伤　伤口一般不宜愈合。一旦有溴沾到皮肤上，先用 20% 的 $Na_2S_2O_3$

溶液冲洗，再用大量水冲洗，用消毒纱布包扎后就医。

（6）吸入刺激性或有毒气体　如吸入氯气、氯化氢气体，可吸入少量酒精和乙醚的混合蒸气解毒。吸入硫化氢或一氧化碳气体而感到不适，应立即到室外呼吸新鲜空气。

（7）毒物进入口内　把 5~10ml 稀硫酸铜溶液加入一杯温水中，内服后用手指伸入咽喉部，促使呕吐，吐出毒物，然后送医院诊治。

（8）触电　立即切断电源，必要时进行人工呼吸并送医院治疗。

（9）起火　立即停止加热、停止通风，关闭电闸，移走一切可燃物，防止火势蔓延。之后要针对起因，选用合适的方法灭火。一般小火可用湿布、石棉或砂土覆盖燃烧物，即可灭火。火势大时可使用泡沫灭火器。电器设备所引起的火灾，只能使用二氧化碳或四氯化碳灭火器灭火，不能使用泡沫灭火器以免触电。有机溶剂（如苯、汽油）或与水能发生剧烈反应的化学药品着火，不能用水灭火，否则会引起更大的火灾，应使用干粉灭火器灭火。

（编写：张莹）

第一章　基本知识和基本操作

一、无机化学实验中常见仪器介绍

无机化学实验中常见仪器见表 1-1。

表 1-1　无机化学实验中常见仪器

仪器名称	规格	主要用途及注意事项
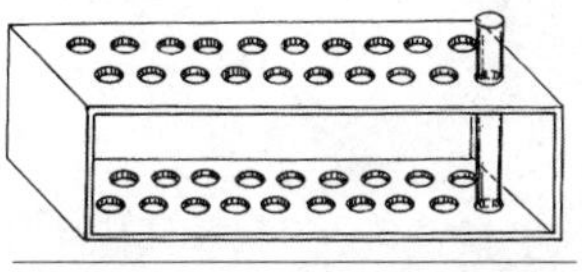 图 1-1　试管和试管架	试管以管口外径（mm）×管长度（mm）表示，分 10×75、10×100、25×150 等规格 试管架分有机玻璃或铝等材质	试管用作简单反应的容器，易于操作、观察，试剂用量少 1. 加热时，用试管夹夹持，管口不能对着有人的方向；为使受热均匀，应移动试管； 2. 反应液体一般不超过容积的 1/2，加热时不超过 1/3 试管架用于放置试管
 图 1-2　离心试管	分有刻度、无刻度两种。以容量（ml）表示，分 5、10、15、25 等规格	用于分离溶液和沉淀； 不能直火加热
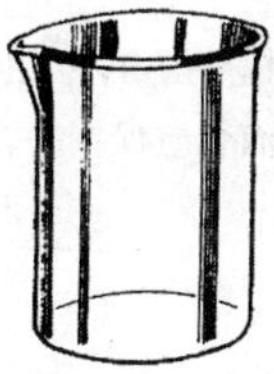 图 1-3　烧杯	以容量（ml）表示，分 50、100、200、500、1000、2000 等规格	用作较多量反应物的反应容器，易于均匀混合；可用于配制溶液或代替水槽 1. 加热时，注意使受热均匀； 2. 用作反应容器时，反应液体一般不宜超过烧杯容积的 1/2
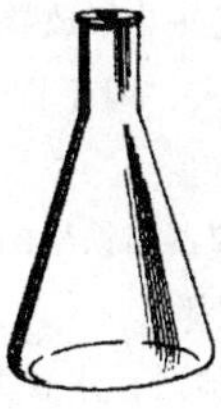 图 1-4　锥形瓶	以容量（ml）表示，分 100、150、200、500 等规格	用作滴定操作或反应容器，因口径较小，便于振荡 1. 加热时，注意使受热均匀； 2. 液体一般不宜超过容积的 1/3

续表

仪器名称	规格	主要用途及注意事项
 图1-5 试剂瓶 （a）广口 （b）细口	分无色和棕色；分细口和广口。以容量（ml）表示，分100、125、250、500、1000等规格	细口瓶用于储存液体试药，广口瓶储存固体试药，棕色瓶存放见光易分解的试药 1. 不能加热； 2. 盛放碱液时，应用橡胶塞或改用塑料瓶存放
 图1-6 滴瓶	分无色和棕色，以容量（ml）表示，分15、30、60等规格	用于盛放少量液体试药，方便取用
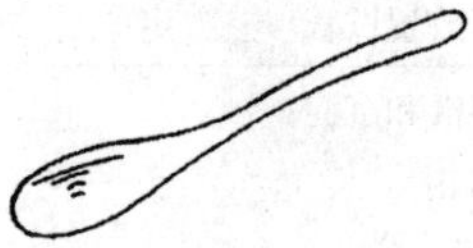 图1-7 药匙	分牛角、瓷、塑料材质	用于取固体药品。对于两端为一大一小的药匙，应根据取用药量的多少选择使用 牛角或塑料药匙不能取灼热药品
 图1-8 毛刷	有多种形状；分大、小、长、短等多种规格	用于洗刷玻璃仪器 应根据待刷玻璃仪器选择合适形状的毛刷
 图1-9 胶头滴管		用于吸取少量液体 1. 避免液体进入橡皮帽内，防止污染； 2. 滴加液体时，滴管应保持垂直，不能触及容器壁
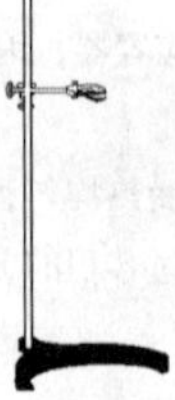 图1-10 铁架台	底座分三角形和长方形，其上配有铁圈和铁夹	用于固定反应容器。铁圈可作为泥三角的支撑架，可替代漏斗架放置漏斗。铁架可安装滴定管夹，用于固定酸碱滴定管 1. 固定仪器时，仪器与铁架台的重心应落在铁架台底盘的中部； 2. 用铁夹固定仪器时，注意力度适中，以防破损仪器

续表

仪器名称	规格	主要用途及注意事项
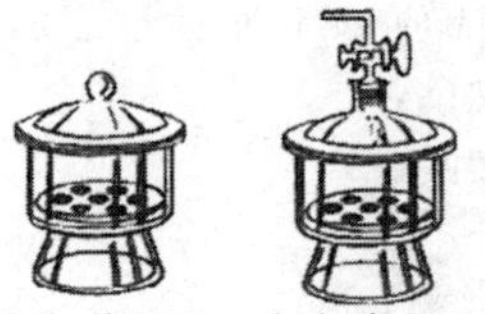 （a）普通型 （b）真空型 图1-11 干燥器	分普通干燥器和真空干燥器。以直径（cm）大小表示	用于干燥药品或仪器 1. 使用前，应在盖子和底座的磨砂部位均匀涂抹真空脂或凡士林。开盖时，应将盖子水平推开，搬动时，应用手指按住盖子，防止滑落； 2. 干燥器中的干燥剂应及时更换； 3. 高温物品应稍冷却后放入
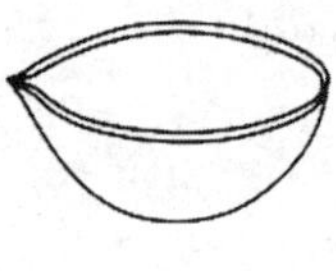 图1-12 蒸发皿	分瓷、石英等材质。规格以直径（mm）表示，分30、40、50、60、85等规格	用于蒸发浓缩液体 1. 根据待蒸发溶液的性质，选择使用不同材质的蒸发皿； 2. 使用时，应避免骤冷
 图1-13 研钵	分瓷、玻璃、玛瑙等材质。以直径（mm）大小表示	用于研磨或混合固体物质；可用作室温固相合成反应的容器 1. 根据固体物质的性质和硬度或实验要求选用不同材质的研钵； 2. 放入量不宜超过研钵容积的1/3； 3. 易爆物质只能轻轻压碎，不能研磨
 图1-14 表面皿	以直径（mm）大小表示，分45、65、75、90等规格	用于盖在烧杯上防止液体迸溅；可进行点滴反应 不能直火加热
 图1-15 容量瓶	分无色和棕色。以刻度以下的容量（ml）表示，分25、50、100等规格	用于准确浓度溶液的配制 1. 不能加热； 2. 磨口与瓶塞要匹配密合
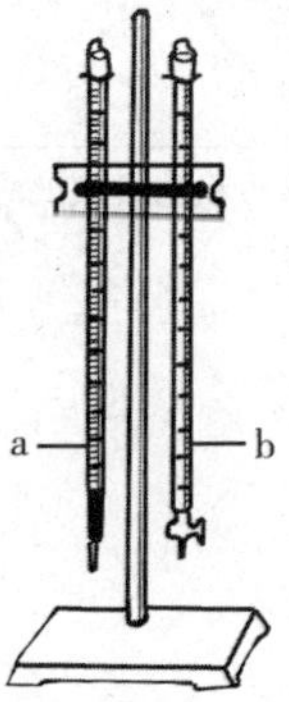 （a）碱式 （b）酸式 图1-16 滴定管	分酸式和碱式；分无色和棕色。以刻度最大标度（ml）表示，分25、50、100等规格	用于滴定分析，可准确读取试液用量；可用于量取准确体积的液体 1. 酸式滴定管与碱式滴定管不能互换使用； 2. 不能加热；不能量取热的液体； 3. 滴定时应先排除尖端部位的气泡； 4. 见光易分解的溶液应使用棕色滴定管； 5. 酸式滴定管可盛装氧化性溶液；碱式滴定管可盛装还原性溶液

续表

仪器名称	规格	主要用途及注意事项
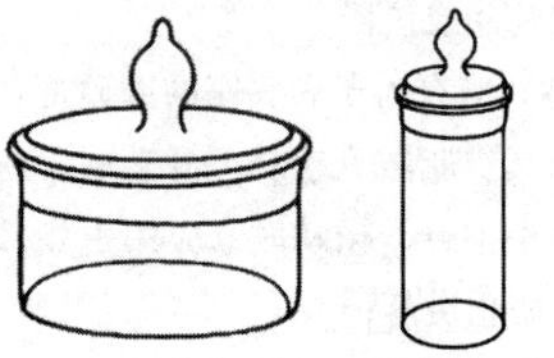 （a）低型　（b）高型 图 1-17　称量瓶	分高型和低型。以容量（ml）表示，分 10、20、25、30、40 等规格	用于固体药品的准确称量 1. 不能加热； 2. 盖与瓶要匹配
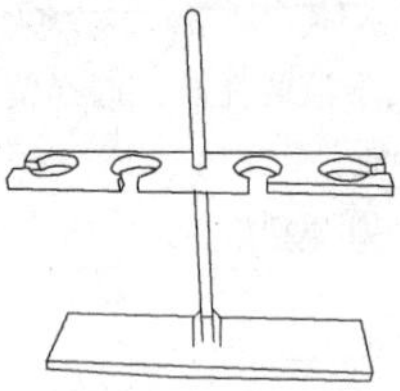 图 1-18　漏斗架	分木质、有机玻璃等材质	过滤时，用于放置漏斗
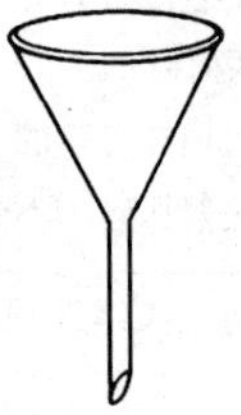 图 1-19　普通漏斗	分长颈和短颈。以直径（mm）表示，分 30、40、60、100 等规格	用于过滤，将固体和液体分离；可用于引导液体或粉末状固体进入小口容器 1. 不能直火加热； 2. 应放在漏斗架上使用，漏斗颈的尖端应靠在盛接液体的容器器壁上
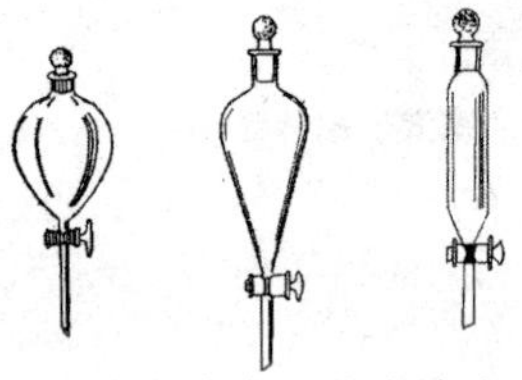 （a）分液　（b）滴液 图 1-20　分液漏斗和滴液漏斗	分液漏斗分球形、梨形等，滴液漏斗为管形。以容量（ml）表示，分 50、100、250、500 等规格	分液漏斗用于分离互不相溶的液体。滴液漏斗用于加入料液 1. 不能加热； 2. 使用前确保活塞处密封； 3. 萃取时，振荡时应注意放气，以免漏斗内压力过大
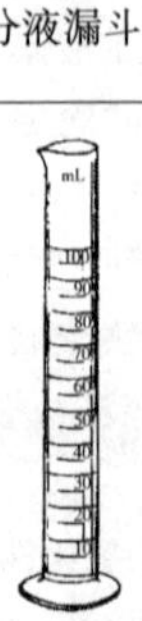 图 1-21　量筒	以最大容量标度（ml）表示，分 5、10、20、25、50、100、200 等规格	用于近似的液体体积的量取 1. 不能作反应容器； 2. 不能加热； 3. 不能配制溶液

续表

仪器名称	规格	主要用途及注意事项
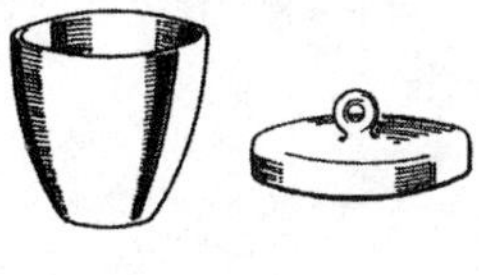 图1-22 坩埚	分瓷、铁、银、镍、铂等材质。以容量（ml）表示规格，分25、30等规格	耐高温，用于灼烧固体 1. 灼烧时，应置于泥三角上； 2. 灼热的坩埚不能骤冷； 3. 灼热的坩埚应放在石棉网上
 图1-23 坩埚钳	分铁、铜、合金等材质制成，表面常镀镍或铬	用于夹取坩埚或坩埚盖 1. 夹取灼热坩埚时，应先预热； 2. 避免接触化学药品，防止腐蚀； 3. 尖端朝上放置以保持清洁
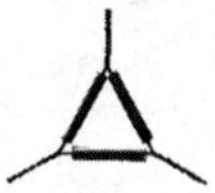 图1-24 泥三角	由铁丝和瓷管制成。以泥三角边长（cm）表示规格	用于盛放坩埚或小蒸发皿。 1. 灼热的泥三角防止骤冷； 2. 选择泥三角时，应使坩埚露出部分不超过其本身高度的1/3
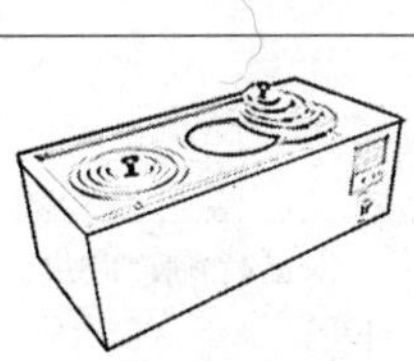 图1-25 恒温水浴		用于低于100℃的恒温控制 1. 加入的水量不应超过容积的2/3； 2. 使用后应及时将水排空，防止生锈
 (a) 布氏漏斗　(b) 吸滤瓶 图1-26 布氏漏斗和吸滤瓶	布氏漏斗为瓷质，以容量（ml）或直径（cm）表示； 吸滤瓶以容量（ml）表示，分50、100、250、500等规格	二者配套使用，进行液体和固体的减压分离 不能加热
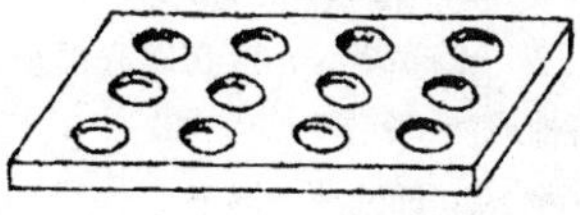 图1-27 点滴板	瓷质，分白色和黑色。以凹穴数表示，有二凹穴、六凹穴等	用于显色反应 白色沉淀用黑色板，有色沉淀用白色板

续表

仪器名称	规格	主要用途及注意事项
 图1-28 洗瓶	塑料材质。以容量（ml）表示，分250、500等规格	盛装纯化水，用于淋洗玻璃仪器、洗涤沉淀等 1. 不能盛自来水； 2. 不能加热
 图1-29 碘量瓶	以容量（ml）表示	用于碘量法或溴酸钾法 滴定时，打开塞子，用纯化水将瓶口及瓶塞上的碘液冲入瓶内
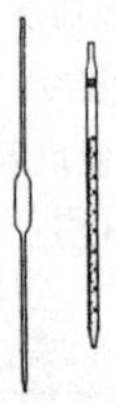 （a）移液管　（b）吸量管 图1-30 移液管和吸量管	移液管为中间膨大形，吸量管为管型。以容量（ml）表示。移液管分10、25、50等规格；吸量管分0.1、0.2、1、2、5、10等规格	用于准确移取体积的液体 不能加热
 图1-31 温度计	分水银温度计和酒精温度计，分别可测最高300℃和100℃的温度	用于测量温度，根据待测物温度范围，选择温度计的种类 1. 热温度计不能骤冷；不能作搅拌棒； 2. 测量温度时，不应使水银球或酒精球触到容器的底部或侧壁
 图1-32 烧瓶	分平底烧瓶和圆底烧瓶两种。以容量（ml）表示，分250、500、1000等规格	用作反应容器，适用于反应物较多，且需要长时间加热的反应。也可以用于蒸馏、回流 1. 加热时，注意受热均匀； 2. 盛放液体的量不宜超过容积的2/3；用作反应容器时，反应液体一般不宜超过烧杯容积的1/3
图1-33 燃烧匙	分铁质或铜质	检验物质可燃性，进行固体燃烧实验 1. 硫磺、钾、钠燃烧时，应在匙底部垫上少许石棉或沙子； 2. 使用后，应立即洗净，擦干

二、常见玻璃仪器的洗涤与干燥

（一）玻璃仪器的洗涤

化学实验经常使用各种玻璃仪器，而所用仪器干净与否直接影响实验结果，为获取准确的实验数据，使用干净的玻璃仪器非常必要。附着仪器上的污物主要有灰尘、可溶性物质、不溶性物质、有机物及油污等。洗涤仪器的方法很多，通常根据实验要求、污物性质及仪器特点来选择合适的洗涤方法。

1. 用水刷洗　用水和毛刷刷洗仪器可以除去灰尘、可溶性物质，以及容易脱落的不溶性物质。

2. 用洗衣粉或合成洗涤剂清洗　由于去污粉中含有碱性物质碳酸钠，而洗衣粉和合成洗涤剂含有表面活性剂，因此它们具有较强的去除不溶性物质、有机物及油污的能力。通常，先用自来水洗刷仪器，再用毛刷沾取少量的洗衣粉水或合成洗涤剂溶液进行刷洗，再用自来水冲洗干净，最后用纯化水润洗3次，纯化水润洗的过程应遵循"少量多次"的原则，通常采用洗瓶，既节约纯化水，又提高润洗效率。

3. 用铬酸洗液洗　对于一些形状特殊的容量仪器，如滴定管、容量瓶、移液管等，不能用毛刷沾取洗衣粉水或合成洗涤剂溶液进行刷洗，常用铬酸洗液洗涤。由于铬酸洗液具有强酸性、强腐蚀性、强氧化性，因此对具有还原性的污物如有机物、油污的去除能力特别强。

用铬酸洗液洗涤仪器时，先用水清洗并尽量把仪器内的残留水倒掉，以免稀释洗液。向仪器中加入少量洗液，倾斜仪器并慢慢转动，使仪器内壁全部被洗液润湿，转动一会儿后，将洗液倒回原洗液瓶中，再用自来水将仪器冲洗干净，最后用纯化水润洗3次。如果先用洗液对仪器浸泡一段时间，或者使用热的洗液，洗涤效果会更好。

铬酸洗液的配制方法是：称取25g重铬酸钾固体，溶于50ml水中，冷却后，向溶液中慢慢加入450ml浓硫酸（注意安全），边加边搅拌，即得铬酸洗液，冷却后贮存在试剂瓶中。

使用铬酸洗液时，应注意以下几点：

（1）用过的洗液应倒回原洗液瓶中，反复使用。

（2）洗液具有很强的吸水性，所以应随时盖严洗液瓶盖。

（3）洗液具有强腐蚀性，因此注意不要溅到皮肤和衣物上，如果不慎溅上，须立即用水冲洗。

（4）当洗液的颜色从暗红变绿时，洗液已经丧失去污能力，不应继续使用。

（5）因铬元素具有毒性，所以应注意废洗液的回收和处理，防止铬污染。

玻璃仪器清洗干净的标准是，仪器用水冲洗后，仪器内壁能被水均匀地润湿，即仪器壁上留有一层薄而均匀的水膜，但不挂水珠。

（二）玻璃仪器的干燥

有些仪器洗涤干净后即可用于实验，但有些实验，特别是要求无水条件下进行的实验，所用的玻璃仪器需要干燥后才能使用。根据具体情况，可以选择合适的干燥方法。

1. 晾干　不急用的仪器在洗净后可以倒置，使其自然干燥。

2. 烤干　急需干燥的仪器，如烧杯和蒸发皿等，可以放在石棉网上用小火烤干。试管可以管口朝下，在小火上移动烘烤，待水珠消失后，再将管口朝上加热，赶尽水汽（图1-34）。

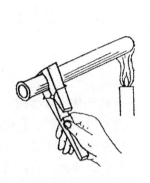 

　　　　图1-34　火烤干　　　　　　　　　图1-35　电吹风吹干

3. 吹干　急需干燥的仪器，可以用电吹风（图1-35）或气流干燥器（图1-36）吹干。

 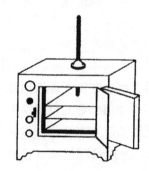

　　图1-36　气流干燥器吹干　　　　　　图1-37　电烘箱烘干

4. 烘干　通常洗净的仪器，可以先将水沥干，然后将仪器口朝下放入电热恒温干燥箱（图1-37）内烘干。

5. 用有机溶剂干燥　在洗净的仪器中，加入少许乙醇或丙酮，转动仪器，使器壁上的水与其互溶，然后倒出混合液，再用电吹风吹干或晾干。

需要特别注意的是，带有刻度的计量仪器，不应该用加热的方法干燥，以免影响仪器的精度。

三、加热与冷却

（一）加热装置

化学实验室中，经常用到酒精灯、煤气灯、电炉及电加热套等加热装置。

1. 酒精灯　酒精灯由灯罩、灯芯和灯壶三部分组成，如图1-38，加热温度通常在400℃～500℃，适用于不需要太高温度的加热实验。点燃酒精灯时，应使用火柴，而不能用燃着的酒精灯，因为洒出的酒精可能引起火灾。向酒精灯内添加酒精时，应先将火焰熄灭，然后借助漏斗添加，最多加入量为灯壶容量的2/3。熄灭时，不能用嘴吹，而应使用灯罩盖，火焰熄灭后，应再提起灯罩一次，防止内部负压影响再次开启。

2. 煤气灯 在有煤气资源的地区，煤气灯是化学实验室最常用的加热装置。

煤气灯有多种样式，但构造原理基本相同，其由灯管和灯座组成（图1-39）。灯管下部有螺旋与灯座相连，并开有作为空气入口的圆孔，旋转灯管可关闭或打开空气入口，以调节空气的进入量；灯座的侧面为煤气入口，用橡皮管与煤气管道相连，灯座的侧面（或下面）有螺旋形的针阀，可调节煤气的进入量。

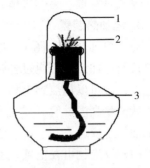

图1-38 酒精灯构造

1. 灯罩；2. 灯芯；3. 灯壶

使用煤气灯时，旋转灯管以关闭空气入口，再旋转针阀，打开煤气入口。在打开煤气管阀门的同时，用燃着的火柴在灯管口处点燃煤气，旋转灯管以导入空气，使煤气完全燃烧，形成正常的蓝色火焰，使用完毕，可以直接将煤气管阀门关闭。

当煤气和空气混合比例合适时，煤气能充分燃烧，获得正常火焰。正常火焰分为三层，如图1-40所示，外层，煤气完全燃烧，称为氧化焰，呈淡紫色；中层，煤气不完全燃烧，称为还原焰，呈淡蓝色；内层，煤气和空气混合并未燃烧，称为焰心，最高温度位于还原焰的顶端与氧化焰之间，温度可达800℃～900℃。

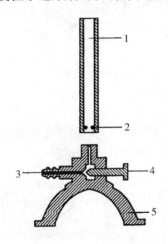

图1-39 煤气灯构造

1. 灯管；2. 空气入口；3. 煤气入口；4. 针阀；5. 灯座

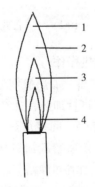

图1-40 正常火焰

1. 氧化焰；2. 最高温处；3. 还原焰；4. 焰心

当空气和煤气的比例不合适时，会产生不正常火焰。若火焰呈黄色，说明煤气没有充分燃烧，应增大空气的进入量。如果空气和煤气的进入量过大时，火焰会脱离灯管而临空燃烧，称为"临空火焰"，这种火焰容易自行熄灭；当煤气的进入量很小，而空气的比例较大时，煤气就会在灯管内燃烧，形成一条细长的绿色火焰，称为"侵入火焰"。遇到上述两种不正常的情况时，应该关闭煤气阀，重新调节后再次点燃。

煤气是有毒、易燃气体，所以实验结束后，应该立即关闭煤气管阀门，保证安全。

3. 电加热装置 实验室常用的电加热装置主要有电炉（图1-41）、电热板（图1-42）、电加热套（图1-43）、管式炉（图1-44）和马弗炉（图1-45）等。电炉、电热板及电加热套可通过外接调压装置控制加热温度。管式炉和马弗炉都属高温加热装置，一般可以加热到1000℃以上，主要用于高温灼烧或高温反应，通常采用热电偶温度计测量高温。管式炉内部为管状炉膛，炉膛中插入一根耐高温的瓷管或石英管，用来抽真空或通入利于反应进行的气体，反应物放入反应舟中，再将其放进瓷管或石英管内，反应物可在空气气氛或其他气氛中进行反应。马弗炉的炉膛为正方形，打开炉门就可放入需要加热的坩埚或其他耐高温容器。如果要灰化滤纸或有机物，在加热过程中应打开几次炉门通入空气。

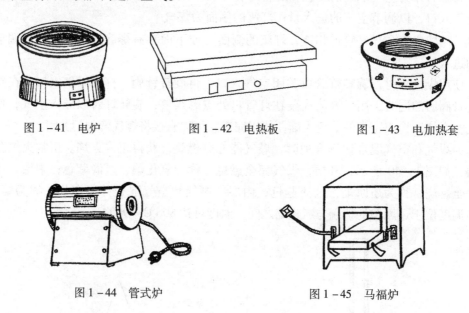

图1-41 电炉　　　图1-42 电热板　　　图1-43 电加热套

图1-44 管式炉　　　图1-45 马弗炉

（二）加热操作

1. 直接加热 加热操作分为直接加热和间接加热两种。直接加热是将被加热物直接放在热源中进行加热，如在煤气灯上加热试管或在马弗炉内加热坩埚。

2. 间接加热 间接加热是先用热源将某些介质加热，介质再将热量传递给被加热物，这种方法叫热浴。常见的方法有水浴、油浴和砂浴等，它们可均匀加热。如果所需的温度不超过100℃，可以采取水浴加热；如果所需温度超过100℃，可以采用油浴或砂浴（图1-46）。油浴即用油代替水浴中的水，一般可达100℃~250℃；砂浴是在铁

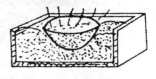

图1-46 砂浴

盘中装入一层均匀的细砂，将需要加热的器皿部分埋入砂中，将铁盘用电炉或煤气灯加热即形成砂浴。测量砂浴温度时，可将温度计埋入器皿附近的细沙中。砂浴的特点是升温较慢，停止加热后，散热也较慢。

3. 液体的加热

（1）加热试管中的液体　加热试管中的液体时，液体的量不应超过试管容积的1/3。应该用试管夹夹持试管，管口稍微向上倾斜（图1-47），注意试管口不能对着有人的方向，以免爆沸喷出的液体伤人。应先加热液体的中上部，再加热底部，并上下移动，使各部分液体受热均匀。

（2）加热烧杯、烧瓶中的液体　加热时应垫上石棉网（图1-48），使仪器受热均匀。加热的液体量不应超过烧杯容积的1/2和烧瓶容积的1/3。烧杯加热时应适当搅拌以免爆沸，烧瓶加热时可视情况加入几粒沸石。

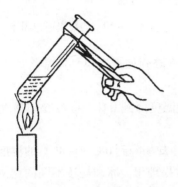

图1-47　加热试管中液体　　　　图1-48　加热烧杯中液体

（3）蒸发浓缩　很多无机化合物的溶解度随温度升高而增大，因此从溶液中将无机物分离出来，可以通过蒸发、浓缩和结晶的步骤完成。蒸发可以除去过多的溶剂，使溶液浓缩达到饱和或过饱和，然后冷却降温使晶体析出。蒸发浓缩通常在蒸发皿中进行，可用电加热套加热，根据产物的性质调节加热的温度，蒸发皿所盛放的液体量不应超过其容积的2/3，注意瓷质蒸发皿不能骤冷，防止炸裂。

4. 固体的加热

（1）加热试管中的固体　在加热试管中的固体时，可用试管夹夹持试管，也可用铁架台和铁夹固定试管（图1-49），管口要略向下倾斜，防止凝结在管口处的水珠倒流至灼热的管底，使试管炸裂。

（2）固体的灼烧　固体需要高温熔融或高温分解或灼烧时，通常在坩埚中进行，加热时，将坩埚置于泥三角上（图1-50），用氧化焰灼烧。先用小火将坩埚均匀加热，再用大火灼烧坩埚的底部。灼烧到符合要求后，停止加热，先在泥三角上稍冷，再用坩埚钳夹持到干燥器内放冷。夹取高温坩埚时，应使用干净的并预热过的干净坩埚钳。坩埚钳使用后，应尖端朝上放置，以保证干净。用煤气灯灼烧的温度可达700℃~800℃，若需要更高温度的灼烧可使用马弗炉，马弗炉可以精确地控制灼烧温度和时间。

（三）冷却方法

有些化学反应需要低温条件下进行，有些反应需要将产生的热量传递出去，制备实验还常从溶液中析出结晶或液态物质凝固等，这些都需要冷却降温。冷却的方法一般是将反应容器置于冷却剂中，通过热量传递达到冷却的目的，有时也可将冷却剂直接加入到反应器中降温。实验室常用如下方法：

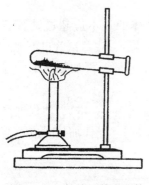
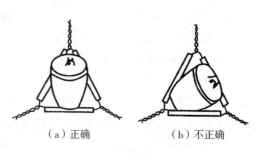

图 1-49　加热试管中固体　　　　图 1-50　灼烧坩埚中固体

1. 水冷却　用水冷却可将被冷却物的温度降到接近室温，通常将被冷却物浸在冷水或在流动的冷水中。

2. 冰水冷却　冰或冰水可将被冷却物的温度降到0℃，将被冷却物直接放在冰或冰水中冷却。

3. 冰盐冷却　冰-无机盐冷却剂可将被冷却物降到0℃～-40℃的温度。制作冰-盐冷却剂时，将盐研细后与粹冰混合，制冷效果理想。冰和盐按不同的比例混合，可以得到不同的冷却温度，见表1-2。干冰-有机溶剂冷却剂可获得-70℃以下的低温。

表 1-2　常见的冷却剂及其最低致冷温度

冷却剂	温度/℃	冷却剂	温度/℃
冰-水	0	$CaCl_2 \cdot 6H_2O$-冰 1:1	-29
NaCl-碎冰 1:3	-20	$CaCl_2 \cdot 6H_2O$-冰 1.25:1	-40
NaCl-碎冰 1:1	-22	液氨	-33
NH_4Cl-碎冰 1:4	-15	干冰	-78
NH_4Cl-碎冰 1:2	-17	液氮	-196

四、试剂的级别及试剂的取用

（一）试剂的级别

按照国家标准，可将化学试剂依据其纯度和杂质含量的高低分成4个等级，参阅表1-3。

表 1-3　化学试剂的级别

级别	名称	英文名称	英文缩写	适用范围	标签颜色
一级品	优级纯	guaranteed reagents	G.R.	精密分析和科学研究	绿
二级品	分析纯	analytical reagents	A.R.	分析和教学工作	红
三级品	化学纯	chemical pure	C.P.	分析和教学工作	蓝
四级品	实验试剂	laboratorial reagents	L.R.	用于一般性的化学实验和教学工作	棕
	生物试剂	biological reagents	B.R.		黄

实验中，应根据不同的实验目的选择不同级别的试剂。化学试剂在分装时，一般把固体试剂装在广口瓶中，把液体试剂放入在细口瓶或带有滴管的滴瓶中，把见光易分解的试剂（如硝酸银等）盛放在棕色瓶内。每个分装的试剂瓶都应贴上标签，标明试剂的名称、浓度、规格等。

（二）试剂的取用

1. 固体试剂的取用 应用干燥洁净的药勺取用，取完试剂后，立即盖好试剂瓶盖。多取的药品不应倒回试剂瓶，防止试剂污染，可将其放在指定的容器中，供他人使用。固体试剂可以用干净的称量纸、表面皿或小烧杯称量。具有腐蚀性、强氧化性或易潮解的固体试剂不能用称量纸，应用称量瓶称量。向试管特别是湿的试管内，加入固体试剂时，可将药勺伸入试管 2/3 处，或将药品放在对折的纸条上，再伸向试管底部，使固体试剂沿着管壁慢慢滑入（图 1-51）。

 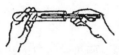 

（a）用药匙向试管中送入固体试剂　（b）用纸槽向试管中送入固体试剂　（c）块状固体沿着试管壁慢慢滑下

图 1-51　取用固体试剂

2. 液体试剂的取用

（1）从滴瓶中取用液体试剂时，应该注意保持滴管垂直，避免倾斜，防止试剂流入橡皮胶头内而将试剂污染。滴加液体试剂时，滴管的尖端不可接触试管或其他容器的内壁，而应在容器口的正上方将液体垂直滴入（图 1-52）。用后的滴管不能放在原滴瓶以外的地方，更不能将其错放入其他滴瓶中。

（2）用倾注法取用液体试剂时，先将瓶盖取下倒放在桌面上，右手握住试剂瓶，有标签的一面朝向手心，将瓶口靠住容器壁，倾斜试剂瓶缓缓倒出液体，使液体沿着干净的玻璃棒或容器内壁流下。取完后，应及时盖好瓶盖，将试剂瓶放回原处。

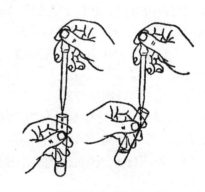

（a）正确　（b）不正确

图 1-52　取用液体试剂

（3）用量筒量取液体时，应左手持量筒，并用大拇指指示所需体积的刻度处，右手持试剂瓶，标签朝手心方向，试剂瓶口应紧靠量筒口的边缘，慢慢注入液体至所需刻度。

五、溶解、结晶和固液分离

（一）固体的溶解

溶解固体时，如固体颗粒较大，可在研钵中研细后再溶解。溶解时常采用搅拌、加热的方法加速溶解。搅拌时，搅拌棒不应触及容器底部和器壁。如需加热，应根据固体物质的热稳定性选择直接加热或水浴加热。在试管中溶解固体时，可用振荡的方法促进溶解，注意不应用手指堵住管口进行振荡。

（二）结晶与重结晶

多数物质的溶液蒸发浓缩到一定程度时，冷却降温就会析出晶体。晶体颗粒的大小与结晶条件有关。如果溶液的浓度较高，溶质在溶剂中的溶解度随温度的降低显著减小时，冷却速度越快，析出的晶体就越细小。反之，越慢析出的晶体就越粗大。搅拌溶液有利于细小晶体的形成，静置溶液有利于粗大晶体的生长。如果溶液的浓度不高时，投入晶种后，自然冷却，静置，会缓慢生成大颗粒晶体。

重结晶是提纯固体物质的重要手段之一。该方法适用于溶解度随温度变化有显著变化的物质。重结晶即是在加热情况下将被纯化的固体溶解于适量的溶剂中，形成饱和溶液，趁热过滤除去不溶性杂质，然后蒸发浓缩、冷却降温，使被纯化的物质结晶析出，过滤使晶体与留在母液中的杂质分离，从而得到较纯净的物质。如果纯度没有达到要求，可再次进行重结晶。

（三）固液分离

固液分离的常见方法有 3 种：倾析法、过滤法和离心分离法。

1. 倾析法　当晶体颗粒较大或沉淀相对密度较大时，静置后它们能很快沉降至容器底部，采用倾析法可将沉淀中的溶液倾入另一容器中，使沉淀与溶液分离，操作方法如图 1-53 所示。如沉淀需要洗涤，可向盛放过滤后的沉淀容器中，加入少量洗涤溶剂，充分搅拌、静置和沉降，再倾出洗涤溶液，如此反复操作几次，即可洗净沉淀，实现固液分离。

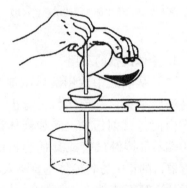

图 1-53　倾析法分离

2. 过滤法　过滤法是最常用的固液分离方法，它包括常压过滤、减压过滤和热过滤 3 种。

（1）常压过滤　当沉淀为胶态或颗粒细小时，用此方法过滤较好，但缺点是过滤速度较慢。

过滤之前，先将滤纸按图 1-54 所示的虚线方向对折 2 次，然后用剪刀剪成扇形，将滤纸打开呈圆锥体形（一侧 3 层，另一侧单层），放入玻璃漏斗中，滤纸的边缘应略低于漏斗边沿 3~5mm（漏斗的角度应是 60°，这样滤纸就能完全贴在漏斗壁上；如果漏斗角度大于或小于 60°，则应适当改变滤纸叠成的角度，使其能与漏斗的角度一致），用手按住滤纸边，再用少量纯化水将其润湿，轻压滤纸四周，赶走滤纸与漏斗壁间的气泡，使其紧贴在漏斗上。

过滤时，将漏斗放在漏斗架上，干净的烧杯放在漏斗下面，调整漏斗架高度，使漏斗尖端紧贴在接收容器内壁，这样可以加快过滤速度，避免溶液溅出（图 1-55）。

用倾析法将溶液沿玻璃棒在三层滤纸一侧缓慢倾入漏斗中，注意液面高度应低于滤纸边缘 2~3mm，使固液分离。如果沉淀需要洗涤，应在溶液转移完毕后，向盛有沉淀的容器中加入少量洗涤溶剂，充分搅拌，待沉淀沉下后，再将上层溶液倒入漏斗，重复洗涤3遍，即可洗净沉淀，注意洗涤沉淀遵从"少量多次"的原则。

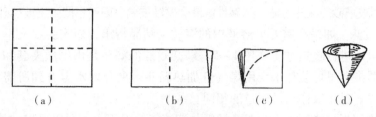

图1-54 滤纸折叠方法

（2）减压过滤 减压过滤即抽滤，可以加快过滤速度，并且获得比较干燥的沉淀。它不适合胶态沉淀和颗粒细小沉淀的过滤，因为胶态沉淀抽滤时会透过滤纸，而细小沉淀会堵塞滤纸的过滤孔因而减慢过滤速度。

减压过滤装置（图1-56）是由布氏漏斗、吸滤瓶、安全瓶及真空泵组成。真空泵可使吸滤瓶内的压力降低，布氏漏斗表面与吸滤瓶内部形成压力差，因而加快了过滤速度。

布氏漏斗是瓷质平底漏斗，中间为多孔瓷板，滤液经过滤纸从小孔流出。选用大小合适的橡皮塞或胶皮垫将布氏漏斗和吸滤瓶连接起来，以保证良好的减压效果；布氏漏斗下端的斜面应正对着吸滤瓶的支管；用耐

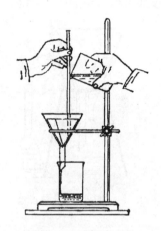

图1-55 常压过滤

压的橡皮管连接吸滤瓶、安全瓶和真空泵；安全瓶可以隔断真空泵和吸滤瓶的直接联系，即使发生倒吸现象也不会污染滤液，若不需要保留滤液时也可以不连接安全瓶。

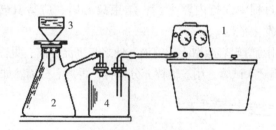

图1-56 减压过滤装置
1. 循环水式多用真空泵；2. 吸滤瓶；3. 布氏漏斗；4. 安全瓶

过滤前，先剪好圆形滤纸，滤纸应比漏斗内径略小，但还须盖住漏斗的全部小孔；用少量水润湿滤纸，打开真空泵，减压使滤纸与漏斗贴紧，然后抽滤。先用倾析法将溶液沿玻璃棒倒入漏斗中，加入量不要超过漏斗容量的2/3，待溶液过滤完毕，再将沉淀转移至漏斗中，待无液滴滴下时，停止抽滤；结束抽滤时，应先将连接吸滤瓶的胶管拔下，再关闭真空泵，以防止倒吸。取下漏斗倒扣在滤纸上，用吸耳球吹漏斗管口，使滤纸和沉淀脱离漏斗，吸滤瓶中的滤液应从吸滤瓶的上口倒出来，而不能从支管倒出。如沉淀需要洗涤，在停止抽气后，用尽量少的溶剂洗涤晶体，使沉淀被溶剂充分

浸润，再减压抽滤，洗涤1～2次即可。

如果过滤的溶液具有强酸性或强氧化性，由于它们能与滤纸发生化学反应而破坏滤纸，所以应该采用玻璃砂芯漏斗过滤。强碱性溶液会腐蚀玻璃，因而不能采用玻璃砂芯漏斗。

（3）热过滤　如果在室温下溶液中的溶质易结晶析出，而实验中又不希望发生这种现象，就可以进行趁热过滤（图1-57）。热过滤的漏斗是由铜质夹套和普通玻璃漏斗组成，铜质夹套里可装热水，用煤气灯加热漏斗，夹套内水温升到所需温度时，即可进行热过滤。过滤操作与常压过滤相同。

热过滤应用折叠滤纸，因为折叠滤纸与溶液接触面积大，所以可以加快过滤速度，折叠滤纸的方法如图1-58所示。

图1-57　热过滤

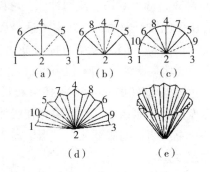

图1-58　热过滤滤纸折叠方法

3. 离心分离法　当被分离的溶液和沉淀数量很少时，采用过滤分离会使沉淀或滤液损失，可以采用离心分离，该方法分离速度快。

离心分离法是将待分离的沉淀和溶液装在离心试管中，放入电动离心机（图1-59）中高速旋转，使沉淀集中在试管底部。离心机在使用时，注意离心试管应对称放置，使离心机在旋转过程中保持内臂平衡；结束离心时，应待其转动完全停止后，再开盖取出试管，以免发生危险。

离心分离后，可用滴管轻轻吸取上层清液，使之与沉淀分离，见图1-60。如果沉淀需要洗涤，可将洗涤溶剂滴入试管，用玻璃棒充分搅拌，再离心分离。如此反复2次即可。

图1-59　电动离心机

图1-60　分离溶液与沉淀

六、固体和液体的干燥

(一) 固体的干燥

固体的干燥可以采用干燥器、挤压或加热等方法，为了提升干燥效率，几种方法还可以配合使用。

1. 干燥器法　干燥器是一种具有磨口盖的厚质玻璃器皿，在磨口上涂有一层薄薄的凡士林，既可防止外界水汽的进入，又能使干燥器具有良好的密闭性。干燥器的底部装有干燥剂（如变色硅胶、无水氯化钙等），中间放置一块多孔瓷板，用以盛放被干燥的物品。打开干燥器时，应左手按住干燥器，右手按住盖的圆顶，沿水平方向移动盖子，如图 1-61 所示。温度很高的物品（如灼烧过恒重的坩埚）放入干燥器后，不能盖严盖子，应留出一条小缝隙，待物品冷却后再盖严，否则容易被里面的热气冲开盖子，或因冷却后干燥器内部产生的负压使盖子难以打开。搬动干燥器时，应用两只手的拇指按住盖子，食指卡住干燥器口的下缘（图 1-62），以防止盖子因滑落而打碎。

干燥器在使用过程中，注意不能存放潮湿物品，以保持干燥；干燥剂的加入量不宜高于底部的 1/2，防止污染存放的物品；干燥剂应及时更换。

图 1-61　开启干燥器

图 1-62　搬动干燥器

干燥器中常用硅胶作干燥剂，硅胶是多孔性物质，吸湿性强。市售的硅胶中常混有氯化钴，无水时为蓝色，吸湿后变成粉红色，粉红色硅胶可用烘箱烘干后，重复使用。除硅胶外，还有无水氯化钙、五氧化二磷、氢氧化钠等可用作干燥剂，实验中根据被干燥物的性质进行选择。

2. 挤压法　在过滤之后，将沉淀夹放在数张滤纸间，通过挤压方式除去溶剂，干燥固体。

3. 加热法　根据被干燥物的热稳定性质，选择合适的加热温度，用烘箱干燥。如果要除去的溶剂是易燃的，不能采用烘箱干燥，可先用电吹风吹走溶剂后，再用烘箱干燥。

(二) 液体的干燥

1. 蒸发或蒸馏法　如果被干燥的液体沸点高难以挥发，则可用蒸发或者蒸馏法除去水分。

2. 干燥器法　当被干燥的液体量很少时，可将其放入干燥器中进行干燥。

3. 干燥剂法　将干燥剂直接加入到被干燥的液体中，常以无水盐作干燥剂，如无水硫酸钠、无水硫酸镁等。加入干燥剂后,应不断进行振荡,加快干燥速度,然后过滤、蒸馏。

七、天平的使用

天平是化学实验中不可缺少的重要的称量仪器。根据实验对质量精确度的要求,可选择不同类型的天平进行称量。常用的天平有托盘天平、分析天平、电子天平,它们都是根据杠杆原理设计而成的。

(一)托盘天平

托盘天平(图1-63)用于一般性称量,精确度不高。

(1)将游码归零,检查指针是否停在刻度盘的中间位置。若不在,可调节托盘下方的平衡调节螺丝,当指针在刻度盘的中间左右摆动大致相等时,天平则处于平衡状态,此时指针指向零点。

(2)称量时,左盘放称量物,右盘放砝码。砝码用镊子夹取,一般5g以内质量可以通过移动标尺上的游码进行添加。当添加砝码后天平的指针停在刻度盘的中间位置时,天平处于平衡状态,此时指针所停的位置称为停点,零点与停点相符时(零点与停点之间允许偏差1小格以内),砝码加游码的质量就是称量物的质量。

(3)称量时应注意以下几点:不能称量热的物品;化学药品不能直接放在托盘上,应根据称量物的具体情况放在干净的表面皿、烧杯或称量纸上;称量完毕,应将砝码放回砝码盒中,将游码拨回"0"位处,将两个托盘放在一侧,以免天平不停摆动。

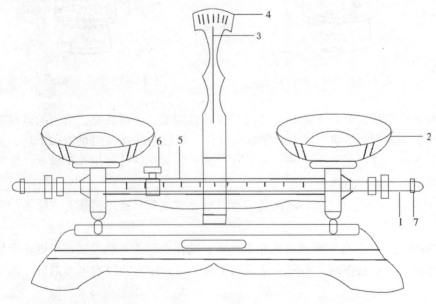

图1-63 托盘天平

1. 横梁;2. 托盘;3. 指针;4. 刻度盘;5. 游码标尺;6. 游码;7. 平衡螺丝

(二)电子天平

电子天平(图1-64)具有称量快捷、使用简单等优点,可自动调零、校准、扣除空白及显示称量结果。

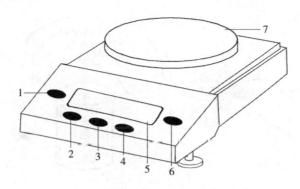

图1-64 电子天平
1. 开关；2. 单位转换键；3. 去皮键；4. 计数键；5. 显示窗；6. 校正键；7. 秤盘

使用方法为：

（1）观察天平水平仪中的水泡是否位于中心，如有偏移，需要调整水平调节螺丝，使天平水平。

（2）接通电源，预热半个小时以上。

（3）轻按ON键，开启显示屏，稍候，出现0.0000g时可以开始称量。

（4）将干净的称量纸（或称量用容器，如小烧杯等）放在称量盘上，关上侧门，轻按TAR键，出现0.0000g时，开启右侧门，向称量纸或称量用容器中加入药品，关上右侧门，显示屏上显示的数值即为被称物的质量。

（5）称量结束后，取走称量纸或称量用容器，关好侧门，轻按OFF键，让天平处于待命状态。再次称量时，轻按ON键就可以继续使用。使用结束后，轻按OFF键，拔下电源插头，罩上天平罩。

（6）使用过程中，注意保持天平清洁，如有药品洒落，应用毛刷及时清理。

（三）称量方法

称取样品一般采用直接法或差减法。

1. 直接法 如果固体样品没有吸湿性，在空气中稳定，就可以采用直接法，即按照上面的电子天平使用方法直接称取样品的质量。

2. 差减法 如果固体样品易吸湿或在空气中不稳定，就可以用差减法称量。先在一个干燥的称量瓶中装入一些样品，在天平上准确称量；将部分样品倒出后，再次准确称量，两次质量之差即为取出样品的质量。称量瓶拿法及从称量瓶中取出样品的方法如图1-65所示。

八、液体体积的量度仪器

（一）量筒

量筒是化学实验中量取液体体积的常用仪器（图1-66），按容量大小有多种规格，可根据需要选择，但注意勿用大容量的量筒量取小体积液体，这样会影响量取的精度。读取数值时，应将量筒放置水平，视线与量筒内液体的弯月面最低处保持水平（图1-

67），然后读取量筒上对应的刻度数值，即为液体体积。

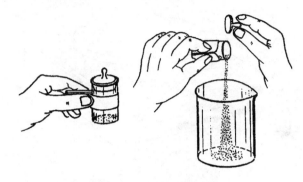

（a）称量瓶的拿法　　（b）从称量瓶中取出样品
图1-65　称量瓶的拿法及从称量瓶中取出样品

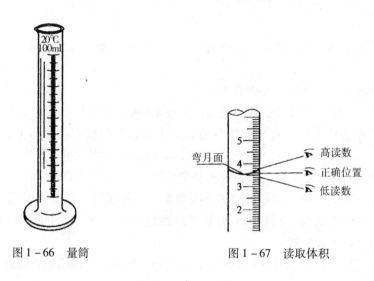

图1-66　量筒　　　　　图1-67　读取体积

（二）滴定管

滴定管是在滴定过程中准确度量滴定溶液体积的玻璃仪器。滴定管一般分为酸式滴定管和碱式滴定管（图1-68）两种，酸式滴定管的刻度管和下端的尖嘴玻璃管通过玻璃旋转塞连接，可盛放酸性溶液和氧化性溶液；碱式滴定管的刻度管和下端的尖嘴玻璃管通过胶皮管相连，管内有玻璃珠控制溶液的流出，碱式滴定管用于盛放碱性溶液和还原性溶液，凡是能与橡胶管发生反应的溶液，如高锰酸钾、硝酸银等都不能装在碱式滴定管中。

滴定管的使用方法如下：

1. 洗涤滴定管　根据滴定管沾污程度，可选用不同的清洗液，如洗洁精水、铬酸洗液等。清洗时在滴定管中加入5~10ml洗液，边转动边将滴定管放平，洗净后将一部分洗液从管口放掉，然后打开旋塞，将剩余的洗液从出口放掉。将洗液彻底放净后，再用自来水冲洗，最后用纯化水洗3次。

碱式滴定管的洗涤方法与酸式滴定管相同，但在用铬酸洗液洗涤时，需注意将玻璃球往上捏，使其紧贴在刻度管的下端，防止洗液腐蚀胶皮管。用自来水和纯化水清

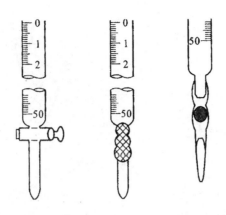

(a) 酸式滴定管　(b) 碱式滴定管

图 1-68　酸式滴定管和碱式滴定管

洗时，应特别注意玻璃球下方死角处的清洗。

2. 涂凡士林　酸式滴定管洗净后，玻璃旋塞处要涂凡士林，起密封和润滑作用。将滴定管平放于实验台上，取下旋塞，用滤纸将旋塞和旋塞套内的水吸干，在旋塞粗端和旋塞套细端分别涂上一薄层凡士林，然后将旋塞插入旋塞套内，单方向旋转旋塞，直到旋塞与旋塞套接触的部位呈透明状态（图 1-69）。涂好凡士林的滴定管应在旋塞末端套上小橡皮圈，以防止旋塞脱落。

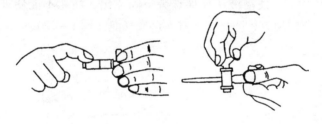

图 1-69　涂凡士林

3. 检漏　向滴定管内装水至最高刻度，垂直夹在滴定管夹上，直立几分钟后观察滴定管口是否有水滴，旋塞与塞槽间隙是否漏水。若不漏，将旋塞旋转180°再观察一次，若漏水则需重涂凡士林。

4. 装入操作溶液　先用操作溶液洗涤滴定管3次，然后将操作溶液装入滴定管至"0.00"刻度以上。装满溶液的滴定管应检查活塞附近或胶皮管内有无气泡，如有气泡，则应排除。酸式滴定管排出气泡的方法为：用右手拿住滴定管使它略为倾斜，左手迅速打开活塞，使溶液快速冲出，将气泡赶掉。碱式滴定管排出气泡的方法为：把胶皮管向上弯曲，玻璃尖嘴斜向上方，用两指捏挤稍高于玻璃球所在处，使溶液从出口喷出，这时一边挤压玻璃球，一边将胶皮管放直，然后再松开手指（图 1-70）。

5. 读数　读数时滴定管需自然下垂，注入或放出溶液后等1~2分钟，待附着在内壁的溶液流下来以后再读数。常量滴定管读数应读到小数点后两位。

读数时用手拿滴定管上部无刻度处，使滴定管保持自然垂直，视线必须与液面保持在同一水平。对于无色或浅色溶液，读其弯月面处最低点所对应的刻度；对于弯月

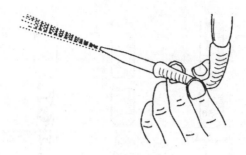

图 1-70 排气泡

面看不清楚的有色溶液如高锰酸钾、碘水等,可读液面两侧的最高点处。无论采用哪种读数方法,都应注意初读数与终读数采用同一标准。

读数时还可以借助读数卡,读数卡就是一张黑纸或深色纸,读数时将它放在滴定管背后,使黑色边缘在弯月下方约1mm,此时看到的弯月面反射层呈黑色,读出黑色弯月面下端最低点的刻度。

6. 滴定操作 使用酸式滴定管时,要用左手控制滴定管活塞,大拇指在管前,食指和中指在管后,无名指和小指向手心弯曲,注意不要使活塞松动,造成漏液。右手持锥形瓶,使滴定管下端进入锥形瓶口约1cm,边滴加溶液边摇动锥形瓶(图1-71)。

使用碱式滴定管时,用左手拇指和食指捏住玻璃球所在处稍高一些的地方,向左边或向右边挤压橡皮管,使溶液从玻璃珠旁空隙流出。注意不能使玻璃珠上下移动,也不要按玻璃珠以下的地方,防止空气进入形成气泡(图1-72)。

   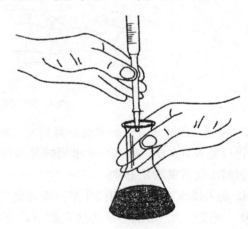

图 1-71 操作酸式滴定管   图 1-72 操作碱式滴定管

滴定通常在锥形瓶中进行,而溴酸钾法、碘量法等需要在碘量瓶中进行。碘量瓶是带有磨口塞和水槽的锥形瓶(图1-29),喇叭形瓶口与瓶塞柄之间形成一圈水槽,槽中加入水就形成水封,可以防止瓶中的 I^- 被空气氧化或 I_2 升华而增大分析误差。反应一段时间后,打开瓶塞,水即流下并可冲洗瓶塞和瓶壁,继续进行滴定。

每次滴定最好都是将溶液装至滴定管"0.00"刻度或稍下一点,以消除上下刻度不均匀而造成的误差。无论使用哪种滴定管,都要掌握好滴液速度(连续滴加、逐滴滴加、半滴滴加），临近终点时，要用纯化水冲洗瓶壁，再继续滴至终点。

（三）容量瓶

容量瓶是一种用于配制具有准确浓度溶液的带磨口塞的仪器，一定温度时刻度线处所标示体积即为规定体积，一般容量为10ml、25ml、50ml、1000ml等规格。

1. 容量瓶的使用 容量瓶使用前应检查是否漏液，方法如下：装水至标线附近，盖好瓶塞，右手托住瓶底边缘，左手食指按住塞子，将其倒立2分钟，观察瓶塞周围是否有水渗出，如果不漏，再把塞子旋转180°，塞紧、倒置，如仍不漏液，则可使用。

2. 溶液的配制 将准确称量的固体物质置于小烧杯中，加入适量水，搅拌使其溶解，将溶液沿玻璃棒转移至容量瓶中（图1-73）；烧杯中的溶液倒尽后，烧杯不应直接离开玻璃棒，而应在烧杯扶正的同时使杯嘴沿玻璃棒上提1~2cm后再离开玻璃棒，如此可避免杯嘴与玻璃棒之间的一滴溶液流到烧杯外面；然后再用少量水冲洗杯壁3~4次，每次的冲洗液按同样的操作转移至容量瓶中。当溶液达到容量瓶的2/3容积时，应将容量瓶沿水平方向摇晃振荡使溶液初步混匀（注意不能倒置容量瓶），然后加纯化水至接近标线，稍等片刻，让附在瓶颈处的水全流入瓶内，再用滴管加纯化水至溶液弯月面最低处与标线相切；盖好塞子，用左手食指按住瓶塞，右手的手指把住瓶底边缘，倒置容量瓶，使瓶内的气泡上升到顶部，边倒转边摇动，如此反复倒转多次，使瓶内溶液充分混合均匀（图1-74）。

图1-73 定量转移溶液

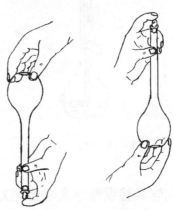

图1-74 混匀溶液

（四）移液管和吸量管

移液管和吸量管（图1-30）用于准确量取一定液体体积的仪器。移液管又称为单标线吸量管，其中间膨大，上下两端为细管状，只能量取某一固定体积（如10.00ml、25.00ml等）的溶液；吸量管又称分度吸量管，用于吸取多种体积的液体，每只移液管和吸量管上都标有使用温度及其容量。

1. 洗涤 用移液管和吸量管吸取液体之前，首先应该用洗液洗净量器内壁，经自来水冲洗和纯化水洗涤3次后，还必须用少量待吸取的溶液润洗内壁3次，以保证溶液的浓度不变。

2. 吸取和放出溶液 用移液管吸取溶液时（图1-75），左手拿洗耳球，右手拇指及中指拿住管颈标线以上的地方，管尖插入液面以下，防止吸空。当溶液上升至标线以上时，迅速用右手食指按紧管口，将管取出液面。左手改拿盛溶液的烧杯，使其倾斜约45°，右手垂直地拿住移液管使管尖紧靠烧杯壁，微微松开右手食指，或用拇指和中指轻轻转动移液管，直到液面缓缓下降到与标线相切时，再次按紧管口，使液体不再流出。放出溶液时，把移液管垂直移入准备接收溶液的容器内壁上方，倾斜容器使它的内壁与移液管的尖端相接触，松开食指让溶液自由流下，待溶液流尽后，再停靠15秒左右，取出移液管。不要把管尖的液体吹出，因为在校准移液管体积时，没把这部分液体算在内（如管上标有"吹"字的移液管，则要将管尖的液体吹出）（图1－76）。

吸量管的使用方法同移液管，只是带有分刻度。

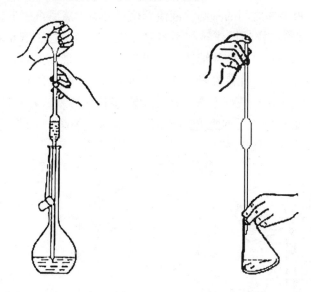

图1－75 移液管吸取液体　　　　图1－76 移液管放出液体

九、化学实验误差与实验数据处理

（一）实验误差

化学是一门实验科学，经常需要定量测定，然后由测量的数据，经过计算得到分析结果。在实验中，同一试样进行重复测定，结果也会不完全一致，即便使用最精密的仪器，也不可能得到完全相同的结果，这说明测量误差是普遍存在的。因此在化学实验中，我们既要正确操作，如实记录，又要树立正确的误差及有效数字的概念，对实验数据进行科学处理，以使分析结果准确。

在定量实验中，会有各种原因导致误差的产生，按照性质不同可以分为系统误差和偶然误差两大类。

1. 系统误差　由实验方法本身的缺陷造成的误差、由仪器或试剂等原因带来的误差（如仪器刻度不够精确，试剂纯度不高等）、由实验者本身的主观原因造成的误差等统称为系统误差。

系统误差通常具有以下特点：
(1) 对测定结果的影响比较稳定。
(2) 使测定结果系统地偏高或偏低。
(3) 在同一条件下重复测定时会重复出现。

2. 偶然误差　某些难以控制的偶然原因（如测定时环境温度、湿度、气压等外界条件的微小变化、由仪器性能的微小波动、实验者偶尔的辨别差异等）造成的误差，称为偶然误差，又称随机误差。

偶然误差难以找到原因，似乎没有规律可言，但它遵守统计和概率理论，可以从多次测量的数据中找到它的规律。一般来说，适当增加测定次数，取多次测定结果的平均值作为分析结果可以减少偶然误差。

除了上述两类误差外，实验者还可能由于疏忽而导致过失误差，如丢失试液、加错试剂、读错数据或计算错误等，过失误差会严重影响分析结果，应避免发生。如果发现过失误差，应立即将实验数据删除。

3. 准确度与精密度

(1) 准确度　准确度指测量值与真实值接近的程度。用误差表示，误差越小，表明测量结果的准确度越高。

误差分为绝对误差和相对误差：

$$绝对误差 = 测量值 - 真实值$$

$$相对误差 = \frac{绝对误差}{真实值} \times 100\%$$

绝对误差可为正值或负值，表示较真实值偏高或偏低，但绝对误差只能显示出误差变化的范围，不能确切地表示测量精度。相对误差表示误差在测量结果中所占百分比，因而准确度常用相对误差表示。

(2) 精密度　精密度是指在相同条件下多次测量结果相接近的程度。精密度用偏差来表示，偏差越小，精密度越高。

在实际工作中，一般要进行多次测定，设一组多次平行测定的数据为：x_1，x_2，$\cdots$ x_n，测量平均值为：

$$\bar{x} = \frac{x_1 + x_2 + \cdots + x_n}{n}$$

式中，$\bar{x}$ 为测量平均值，n 为测量次数。

偏差分为绝对偏差、相对偏差、平均偏差、相对平均偏差、方差、标准偏差及相对标准偏差，各种偏差的计算公式为：

$$绝对偏差 = d_i = x_i - \bar{x}$$

$$相对偏差 = \frac{d_i}{\bar{x}} = \frac{x_i - \bar{x}}{\bar{x}} \times 100\%$$

$$平均偏差 = \bar{d} = \frac{|d_1| + |d_2| + \cdots + |d_n|}{n}$$

$$相对平均偏差 = \frac{\bar{d_i}}{\bar{x}} \times 100\%$$

$$\text{方差} = \frac{\sum_{i=1}^{n}(x_i - \bar{x})^2}{n-1}$$

$$\text{标准偏差} = s = \sqrt{\frac{\sum_{i=1}^{n}(x_i - \bar{x})^2}{n-1}}$$

$$\text{相对标准偏差} = \text{RSD}\% = \frac{s}{\bar{x}} \times 100\%$$

(二) 有效数字

1. 有效数字的定义及记录方法 有效数字为实验中实际测得的数字,它包括几位准确的可靠数字和最后一位不够准确的估读数字。

实验中采用几位有效数字,应根据测量仪器和观测的精确程度进行决定。例如用托盘天平称量某物质的质量为 6.8g,因为托盘天平可以精确到 0.1g,所以该物质的质量可以表示为 (6.8g ± 0.1g),它具有 2 位有效数字;如果改用分析天平称量,其结果为 6.8125g,由于分析天平能精确到 0.0001g,所以该物质质量可以表示为 (6.8125g ± 0.0001g),它具有 5 位有效数字。

有效数字的位数与仪器的精密程度有关,在实验中,超过或低于仪器精密程度的有效数字都是不恰当的。在托盘天平上称得 6.8g,不能记为 6.8000g,因其夸大了托盘天平的精密程度;在分析天平上称得 6.8000g,也不能记为 6.8g,因其降低了分析天平的精密程度。

有效数字的位数可用以下例子说明:

数值: 0.0035   0.0305   45.00   4.5000
有效数字位数: 2   3   4   5

在有效数字中,0 需要特别注意,它有时代表有效数字,有时则不代表,应视其在数值中的位置而定。0 在数字前面,它仅起定位作用,不算有效数字,如 0.0035 的有效数字是 2 位;0 在数字中间,是有效数字,如 4.003 的有效数字是 4 位;0 在数字后,是有效数字,如 4.50 的有效数字是 3 位。

以 0 结尾的正整数,有效数字不确切,如 2500,应根据实际情况改写,有效数字是 2 位时应写成 2.5×10^3;有效数字是 3 位时,应写成 2.50×10^3。

pH、lgK 等对数的有效数字的位数取决于小数部分的数字位数。如 pH = 10.20,其有效数字位数应为 2 位,这是因为该数据由 $c(H^+) = 6.3 \times 10^{11}$ mol/L 而得来。

2. 数字的修约 在实验结果处理过程中,所涉及的测量值的有效数字位数可能不同,因此需要先按照一定的运算规则,确定各测量值的有效数字位数,然后将后面多余的数字舍弃,这个过程称为"数字的修约"。数字的修约一般采用"四舍六入五成双"的规则。

当末位有效数字后面的第一位数字小于或等于 4 时则舍去;当大于或等于 6 时则进位;当等于 5 时,则看末位有效数字是奇数还是偶数决定取舍,是奇数则进位,是偶数则舍去。

例如取四位有效数字,24.0248 应为 24.02;27.327 应为 27.33;23.025 应为

23.02；23.055 应为 23.06。

3. 有效数字的运算规则

（1）加减运算　在进行加减运算时，所得结果的小数点后面的位数应与各加减数中小数点后面位数最少者相同。如：

$$15.34 + 0.00850 + 1.434 = 16.78$$

（2）乘除运算　在进行乘除运算时，所得结果的有效数字的位数应与各数中最少的有效数字位数相同，而与小数点的位置无关。如：

$$2.3 \times 0.524 = 1.2$$
$$25.64 \div 1.05782 = 24.24$$

（3）对数运算　在对数运算中，真数有效数字的位数应与对数尾数的位数相同。如：

$$\lg 15.36 = 1.1864$$

对数的整数是用于定位的，不是有效数字，因此上面的计算不能写成 $\lg 15.36 = 1.186$。同样当 pH = 5.28 时，$C_{H^+} = 5.2 \times 10^{-6}$ mol/L，不能写成 $C_{H^+} = 5.20 \times 10^{-6}$ mol/L。

在计算中，为简便起见，可以在运算前先进行数字的修约，再进行计算。

（三）实验数据表达与处理

化学实验数据的表达与处理方法主要有列表法、图解法和数学方程式 3 种，我们主要介绍常用的列表法和图解法。

1. 列表法　将实验数据列入简明合理的表格中，使得全部数据一目了然，可以起到化繁为简的作用，有利于观察实验结果，分析和阐明实验结果的规律性。列表处理实验数据时应注意以下几点：

（1）每个表必须有简明达意的名称。

（2）表格的横排为行，纵排为列，每个变量占一行或一列，第一栏要写清变量的名称和量纲。

（3）表中的数据应注意有效数字的位数，同一列数字的小数点应对齐。

2. 图解法　对实验数据用图解法来处理，即用一种线图来描述所研究的变量间的关系。图解法能直接显示数据的特点、数据的变化规律，还能由线图求算变量的中间值，确定经验方程中的常数，如求斜率、截距、外推值及内插值等。下面简要介绍用直角坐标纸作图的要点：

（1）选取坐标轴　在坐标纸上画两条互相垂直的直线，一个为横轴一个为纵轴，分别代表实验数据的两个变量。习惯上以自变量为横坐标，因变量为纵坐标，坐标轴旁需要标明代表的变量和单位。

（2）坐标轴比例选择　坐标纸标度应能表示出全部有效数字，从而使图中得到的精密度与测量得到的精密度相当。考虑到图的大小布局能使数据的点分散开，不一定所有的图都把 0 作为坐标原点。

（3）标定坐标点　根据数据的两个变量在坐标内确定坐标点，同一曲线上各个相应的坐标点要用同一种符号表示，不同的曲线需用不同的符号表示。

(4) 画出曲线　用均匀光滑的曲线（或直线）连接坐标点，曲线（或直线）描述了坐标点的变化情况，不必要求它们通过全部坐标点，但是应使坐标点均匀分布在靠近曲线（或直线）的两边。

十、试纸的使用

（一）石蕊试纸和 pH 试纸

石蕊试纸和 pH 试纸都用以检验溶液的酸碱性，pH 试纸还可检查溶液的酸碱度。使用石蕊试纸时，可先将石蕊试纸剪成小块，放在干燥清洁的点滴板或表面皿上，再用玻璃棒蘸取待测的溶液，滴在试纸上，于半分钟内观察试纸的颜色变化（酸性显红色，碱性显蓝色）。不得将试纸浸入溶液中进行检测。检查挥发性物质的酸碱性时，可将石蕊试纸用水润湿，然后悬放在气体出口处，观察试纸颜色变化。

pH 试纸有两种，一种是广泛 pH 试纸，pH 变色范围 1~14，用以粗略检验液体的 pH 值。另一种是精密 pH 试纸，用于较精细溶液 pH 值的测定，它的种类很多，可根据不同的要求选择。

使用 pH 试纸的方法与石蕊试纸基本相同。

试纸应密闭保存，用镊子取用，以免污染变色。

（二）醋酸铅试纸

用以定性检验反应是否有 H_2S 气体产生。此试纸用醋酸铅溶液浸泡过，使用时应先用纯化水润湿，放在试管口，将待测溶液酸化，如有 S^{2-} 离子，则生成的 H_2S 气体遇到试纸生成 PbS 黑色沉淀，而使试纸呈黑褐色并有金属光泽。若溶液中 S^{2-} 离子浓度太低，则不易检出。

（三）淀粉－碘化钾试纸

用以定性检验氧化性气体（Cl_2、Br_2 等）。试纸用淀粉－碘化钾溶液浸泡过，使用时用纯化水润湿并放在试管口上，氧化性气体遇到试纸，将 I^- 氧化为 I_2，I_2 遇到试纸上的淀粉而使试纸变蓝。有时试纸变蓝后有褪色现象发生，这是由于气体氧化性太强的原因，导致 I_2 进一步氧化成 IO_3^-。

（编写：刘迎春）

第二章　基本操作训练实验

Chapter 2　Basic Operation Training Experiments

实验一　称量练习

【预习内容】
1. 电子天平的使用方法。
2. 直接称量法和减量称量法。

【思考题】
1. 如何选择合适的电子天平？使用环境有何要求？
2. 试样的称量方法有几种？如何选择合适的称量方法？
3. 在减量称量法中，若称量瓶内的试样吸湿，对称量结果有无影响？若试样倒入锥形瓶后吸湿，对称量结果有无影响？

【实验原理】

电子天平是最新一代的天平，它的支撑点采取弹簧片代替机械天平的的玛瑙刀口，用差动变压器取代升降枢装置，它是根据电磁力平衡原理，直接称量，全量程不需要砝码，放上被测物质后，在几秒钟内达到平衡，直接显示读数，因此具有体积小、使用寿命长、性能稳定、称量速度快、精度高的特点。此外，电子天平还具有自动校正、自动去皮、超载显示、故障报警等功能，以及具有质量电信号输出功能，且可与打印机计算机联用，进一步扩展其功能，如统计称量的最大值、最小值、平均值和标准偏差等。由于电子天平具有机械天平无法比拟的优点，尽管其价格偏高，但也越来越广泛的应用于各个领域，并逐步取代机械天平。

【实验仪器与药品】

电子天平，干燥器，称量瓶，毛刷。
K_2CrO_4（固体）。

【实验内容】

1. 称量前的检查
（1）检查天平盘内是否干净，必要的话予以清理。
（2）检查天平是否水平，若不水平，调节底座螺丝，使气泡位于水平仪中心。
2. 仪器预热
（1）仪器开机后灯及电子部件需预热平衡，故开机预热 30 分钟才能进行测定工作，才能保证电子天平的稳定及称量数据的准确。

（2）关好天平门，轻按 ON 键，LTD 指示灯全亮，松开手，天平先显示型号，稍后显示为零点，即可开始使用（注意选择零点制式）。

3. 直接称量法

（1）关好天平门，按 TAR 键清零。

（2）打开天平左门，将称量瓶放入托盘中央，关闭天平门，待稳定后读数。

（3）记录后打开左门，取出称量瓶，关好天平门。

4. 增量法

（1）本操作可以在天平中进行，用左手手指轻击右手腕部，将牛角匙中样品慢慢震落于容器内，当达到所需质量时停止加样，关上天平门，显示平衡后即可记录所称取试样的质量。

（2）打开左门，取出容器，关好天平门。

增量法要求称量精度在 0.1mg 以内。如称取 0.5000g 石英砂，则允许质量的范围是 0.4999~0.5001g。超出这个范围的样品均不合格。若加入量超出，则需重称试样，已用试样必须弃去，不能放回到试剂瓶中。操作中不能将试剂撒落到容器以外的地方。称好的试剂必须定量的转入接收器中，不能有遗漏。

5. 减量法称量操作

（1）按 TAR 键清零。

（2）将装有试样的称量瓶放置天平盘中央，关闭天平门（左进左出）。

（3）待天平数据稳定后，及时准确记录装有试样的称量瓶的质量，记作 M_0。

（4）取出称量瓶，悬在容器上方，使称量瓶倾斜，打开称量瓶盖，用盖轻轻敲瓶口上缘，渐渐倾出样品。

（5）估计倾出的试样接近所需要的质量时，慢慢地将瓶竖起，再用称量瓶盖轻敲瓶口上部，使粘在瓶口的试样流回瓶内，盖好瓶盖，将称量瓶放回天平盘上，再准确称其质量，记作 M_1。

（6）两次称量之差（$M_0 - M_1$）即为倒入容器里的试样质量，计算倒出的试样量。

6. 称量工作结束后，取出称量瓶，放回干燥器内，天平回零，关闭电源，清理工作台。

Experiment 1 Weighing Exercise

Preview

1. Usage of electronic balance.

2. Direct weighing method and decrement weighing method.

Questions

1. How can the appropriate electronic balance be selected? Are there any requirements for application environments?

2. How many methods of weighing samples are there?

3. Is there an influence on weighing results if the samples in weighing bottles absorb mois-

ture with decrement weighing method? Is there an influence on weighing results if the samples absorb moisture in conical flasks?

Principles

An electronic balance is the latest generation of balance, which substitutes leaf springs for agate knife edges of the mechanical balance as the supporting point, differential transformers for lifting devices. According to the electromagnetic force balance principle, it can be used for direct weighing without weights within full range. After placing measured matter on it, it can achieve balance within several seconds and show reading directly. So it is characterized by its small size, long service life, stable performance, fast weighing speed and high precision. Moreover, the electronic balance has many functions including automatic calibration, automatic peeling, overload display, failure alarm and so forth. In addition, it has the output function of the quality electric signals, and can be combined with the printer and computer so as to extend its functionality further, such as counting the maximums, minimums, averages and standard deviations of weighing. Although its price is a bit high, electronic balance has incomparable advantages of mechanical balance, so its applications are wider and wider in various fields, and it is replacing the mechanical balance gradually.

Instruments and Chemicals

electronic balance, dryer, weighing bottle, brush.

potassium chromate.

Procedures

1. Examination before weighing

(1) Check if the balance pan is clean and it should be cleaned if necessary.

(2) Check if the balance is horizontal. If not, adjust the base screw and locate the bubble in the center of level.

2. Pre-heating

(1) When the equipment begins to run, lights and electronic parts need to achieve pre-heating balance, so assessment work has to be performed after 30-minute preheating in order to ensure the stability of electronic balance and accuracy of weighing data.

(2) Close the door of the balance; press the key ON softly; loose hands when all the LTD indicator lights flash on; after electronic balance showing model number and then zero, it can be used. (Pay attention to the model of Zero.)

3. Direct weighing method

(1) Close the door of the balance, and press the key TAR for zero clearing.

(2) Open the left door of the balance, place the weighing bottle on the center of the pan, close the door of the balance, read when it is stable.

(3) After reading, open the left door, take the weighing bottle out and then close the door of the balance.

4. Increment method

(1) This operation can be performed in a balance. First, use left-hand fingers to tap the right wrist; and then, shake off the samples in the horn medicine spoon into the vessel and stop adding until reach the weight needed; at last, close the door of the balance, record the quality of the weighed sample after displaying balance.

(2) Open the left door, take out the weighing bottle and close the door of the balance.

The increment method needs controlling the weighing accuracy within 0.1 mg. For example, if 0.5000 g quartz sand is weighed, the range of the weight allowed is from 0.4999 g to 0.5001 g. The samples beyond this range are not qualified. If add more, then the sample should be weighed once more; the samples used must be rejected, and cannot be put back into the reagent bottle. During the operation, the reagents cannot be dropped out of the vessel. The samples that have been weighed well must be transferred into receivers without any omission.

5. Operations of decrement weighing method

(1) Press the key TAR for zero clearing.

(2) Place the weighing bottle of samples onto the center of the balance pan, close the door of the balance (left in and left out).

(3) Till the balance is stable, timely and accurately record the weight of the weighing bottle of samples as M_0.

(4) Take the weighing bottle out, slant it over the vessel, open its cap and tap the surface of the bottle mouth with the cap and decant the sample out little by little.

(5) When the weight of the sample poured is close to the needed value, erect the bottle slowly and then tap the upper part of the bottle mouth with the cap of the weighing bottle in order to let the sample sticking to the mouth of the bottle flow back into the bottle. Put the cap back on the bottle, place the weighing bottle back onto the pan of the balance, weigh its weight again and record it as M_1.

(6) Difference between the two weighing values ($M_0 - M_1$) is the weight of the sample that is poured into the receiver. And calculate the sample weight.

6. After weighing, take the weighing bottle out, place it back to the dryer, till the balance returning to zero, turn off the power and clean the worktable.

Vocabulary

electronic balance	电子天平	conical flasks	锥形瓶
weighing bottle	称量瓶	dryer	干燥器
potassium chromate	K_2CrO_4	quartz sand	石英砂

(编写:张莹,英文核审:肖琰)

实验二 容量仪器的校正

【预习内容】
1. 滴定管、移液管和容量瓶的使用方法。
2. 容量仪器校准的意义。

【思考题】
1. 为什么要进行容量仪器的校正？影响容量仪器校正的因素有哪些？
2. 将水从滴定管中放入容量瓶中时应注意什么？
3. 称量水的质量时为什么只需精确至 0.01g？

【实验原理】

滴定管、移液管和容量瓶是滴定分析中所使用的主要容量仪器，严格讲容器的体积不一定与其标示值完全相符，因此，在准确度要求很高的实验中，必须进行容量仪器的校准。

由于玻璃具有热胀冷缩的特性，不同温度下玻璃仪器的容积也略有不同，因此，校准玻璃仪器时选择 20℃ 为标准温度，即将玻璃仪器校准到 20℃ 时的实际容积。

常用的校准方法为衡量法，又叫称量法，在分析天平上称出容器所容纳或放出的水的质量，再根据水的密度换算出在 20℃ 时的标准容积即为该容器的实际容积。换算时要考虑以下几个方面的影响：
1. 温度对水的密度的影响。
2. 温度对玻璃仪器膨胀系数的影响。
3. 在空气中称量时空气浮力的影响。

综合考虑上述三种影响因素，经总校准后不同温度下纯水的密度见表 2-1。

表 2-1 不同温度下纯水的密度

温度（℃）	密度（g/ml）	温度（℃）	密度（g/ml）
10	0.9984	21	0.9970
11	0.9983	22	0.9968
12	0.9982	23	0.9966
13	0.9981	24	0.9964
14	0.9980	25	0.9961
15	0.9979	26	0.9959
16	0.9978	27	0.9956
17	0.9976	28	0.9954
18	0.9975	29	0.9951
19	0.9973	30	0.9948
20	0.9972		

根据此表可以校准容量仪器。例如 21℃ 时由滴定管放出 10.10ml 的水，称得其质量为 10.08g，查表得 21℃ 水的密度 0.9970g/ml，因此，其实际容积为：

$$\frac{10.08}{0.9970} = 10.11 \text{ (ml)}$$

在有些实验中，并不需要校准容量仪器，只需要确定某些容量仪器间的相互关系，这时可以采用相对校准方法。如用25ml移液管吸取纯化水10次，放入250ml的容量瓶中，观察液面弧线下缘是否恰在标线刻度处，如果不在，需要做新的标记。

【仪器与试剂】

容量瓶（50ml、250ml），滴定管（50ml），移液管（25ml）。

【操作步骤】

1. 酸式滴定管的校正　先将一个50ml容量瓶洗干净，再将其外部擦干，准确称量容量瓶质量（精确至小数点后两位，即0.01g）。将酸式滴定管洗干净后加入纯化水，调节刻度至"0.00"处，记录此时水的温度，然后以每分钟约10ml的速度放出10ml水（误差允许在0.1ml范围内），置于已称重过的容量瓶中，盖上容量瓶盖，再称出其质量，两次质量之差即为放出水的质量。按照同样的方法称量滴定管中从10～20ml、20～30ml等刻度间水的质量。最后用每次得到的水的质量除以该实验温度时水的密度，就可以得到滴定管各部分水的实际容积。表2-2为滴定管校正示例。

表2-2　滴定管校正示例

（25℃水的密度0.9961g/ml）

滴定管读数	容积（ml）	瓶加水的质量（g）	水的质量（g）	实际容积（ml）	校正值	总校正值
0.02		29.20				
10.12	10.10	39.28	10.08	10.12	0.02	0.02
20.09	9.97	49.19	9.91	9.95	-0.02	0.00
30.07	9.98	59.18	9.99	10.03	0.06	0.06
40.02	9.95	69.13	9.95	9.99	0.04	0.10
49.96	9.94	79.01	9.88	9.92	-0.02	0.08

2. 容量瓶的校正　将容量瓶洗干净，自然晾干后称其质量。然后注入纯化水至刻度线，再次称量其质量，两次称量质量之差即为水的质量，然后查表计算容量瓶的实际容积。

3. 移液管和容量瓶的相对校正　用一只洗净的25ml移液管吸取纯化水10次，放入洁净并干燥的250ml的容量瓶中，观察液面弧线下缘是否恰在标线刻度处，如果不在，做新的标记。经相互校准的移液管和容量瓶可以配套使用。

Experiment 2　Calibration of Volumetric Glassware

Preview

1. How to use buret, volumetric pipet and volumetric flask.

2. Calibration of volumetric glassware.

Questions

1. Why is it necessary to calibrate the volumetric glassware before use? What factors influence the calibration of volumetric glassware?

2. What precautions should be taken when draining the water in buret into a container?

3. When measuring the weight of water, you only need it accurate to 0.01g. Why?

Principles

Buret, volumetric pipet and volumetric flask are typically used glassware in titrimetric analysis. Strictly speaking, it is never safe to assume that the volume delivered by or contained in any volumetric instrument is exactly the amount indicated by the calibration mark. Therefore, in precise work, calibration must be performed before use.

Because of the thermal expansion, the volume of volumetric glassware varies with temperature. 20℃ has been chosen as the normal temperature for calibration of much volumetric glassware. The device should be calibrated to the volume it holds at 20℃.

Gravimetric method is usually used to calibrate volumetric glassware. The mass of water that is contained in or delivered by the device are measured on analytical balance. This mass data is then converted to volume data using the tabulated density of water at the temperature of calibration. The following issues should be concerned during the mass-volume conversion.

(1) the influence of temperature on the density of water.

(2) the influence of temperature on the thermal expansion coefficient of glassware.

(3) the influence of the buoyancy effects of air during gravimetric analysis.

Taken together, densities of water at different temperature are corrected and tabulated in Table 2 – 1.

Table 2 – 1   Density of Water

Temperature (℃)	Density (g/ml)	Temperature (℃)	Density (g/ml)
10	0.9984	21	0.9970
11	0.9983	22	0.9968
12	0.9982	23	0.9966
13	0.9981	24	0.9964
14	0.9980	25	0.9961
15	0.9979	26	0.9959
16	0.9978	27	0.9956
17	0.9976	28	0.9954
18	0.9975	29	0.9951
19	0.9973	30	0.9948
20	0.9972		

The table can be used for the calibration of volumetric glassware. For instance, a buret delivered 10.10ml of water with a mass of 10.08g at 21℃. As shown in Table 2 – 1, water has a density of 0.9970g/ml at this temperature. Therefore, the corrected volume of this buret can be calculated as follow.

$$\frac{10.08}{0.9970} = 10.11(\text{ml})$$

In some cases, it is required only to know the relationship between the volumes of the two items of glassware used in the analysis without knowing the absolute volume of either one. This is called relative calibration. For example, pipet 25ml purified water 10 times into a 250ml volumetric flask with a 25ml volumetric pipet, and check whether the meniscus of liquid is at the same level with the indicated level of the flask. If they appear at different levels, the flask should be corrected.

Instruments and Chemicals

volumetric flasks (50ml, 250ml), buret (50ml), volumetric pipet (25ml).

Procedures

1. Calibration of an acid buret

Clean a 50ml volumetric flask and then wipe the surface dry. Weigh a volumetric flask on an analytical balance, and make the reading to a precision of 0.01g. Fill the cleaned acid buret with purified water. Make sure the tip is free of bubbles. Drain into a waste beaker until it is at, or just below, the zero mark. Measure the temperature of the water used. Deliver 10ml (accurate to 0.1ml) of water from the buret into the volumetric flask at a speed of 10ml per minute. Then re-weigh the flask. Calculate the difference of the two mass values. Repeat the process for 10—20ml, 20—30ml of water delivered. Calculate the true volume delivered by dividing the masses of water respectively with its density at the experimental temperature. An example for calibrating a buret was shown in Table 2-2.

Table 2-2 Calibration of a Buret at 25℃

(Density of water at 25℃:0.9961g/ml)

Buret reading	Volume delivered (ml)	Mass of flask plus water (g)	Mass of water delivered (g)	Actual Volume delivered (ml)	Correction	Total correction
0.02		29.20				
10.12	10.10	39.28	10.08	10.12	0.02	0.02
20.09	9.97	49.19	9.91	9.95	-0.02	0.00
30.07	9.98	59.18	9.99	10.03	0.06	0.06
40.02	9.95	69.13	9.95	9.99	0.04	0.10
49.96	9.94	79.01	9.88	9.92	-0.02	0.08

2. Calibration of a volumetric flask

Clean a flask and allow it to air dry. Weigh the mass of the flask on an analytical balance. Fill the flask with purified water to the mark and re-weigh it. Calculate the true volume of the flask using the method outlined above.

3. Relative calibration of volumetric pipet and flask

Pipet 25ml of purified water into a clean and dried 250ml volumetric flask with a 25ml volumetric pipet. Repeat the process 9 more times. Check whether the meniscus of water is at the mark of the flask. If not, remark the flask. The two mutually calibrated items can therefore be used in combination of each other.

（编写：凌俊红，英文核审：肖琰）

实验三　酸碱滴定练习

【预习内容】

1. 酸碱滴定的原理。
2. 酸碱滴定管的操作。
3. 酸碱滴定终点的判断。

【思考题】

1. 如何排除碱式滴定管中的气泡？
2. 用纯化水洗过的滴定管为什么还要用待装溶液洗涤3次？
3. 滴定中所用的锥形瓶是否需用标准溶液洗涤或干燥处理？
4. 在标定 NaOH 溶液浓度的过程中，以下情况对结果有何影响？
（1）滴定管中有气泡。
（2）快到终点时没有用洗瓶冲洗锥形瓶内壁。
（3）滴定结束时滴定管尖嘴处挂有液滴。

【实验原理】

酸碱中和反应的实质是：

$$H^+ + OH^- = H_2O$$

当反应达到化学计量点时，酸碱物质量的关系为：

$$C_{酸}V_{酸} = C_{碱}V_{碱}$$

式中，$C_{酸}$ 为酸的浓度；$C_{碱}$ 为碱的浓度；$V_{酸}$ 为酸的体积；$V_{碱}$ 为碱的体积。

因此酸碱溶液通过滴定，可以确定它们中和时所需的体积比，从而确定它们的浓度比，如果其中某一溶液的浓度已知，就可以求出另一溶液的浓度。

酸碱滴定的终点可以借助指示剂颜色的变化来确定，酸碱指示剂在不同的 pH 范围内可以显示不同的颜色。酚酞在 pH 大于10.0 时显红色，在 pH 小于8.0 时无色，在 pH 8.0~10.0 时为其变色范围；甲基红在 pH 小于4.4 时显红色，在 pH 大于6.2 时显黄色，在 pH 4.4~6.2 时为其变色范围。滴定操作中可以根据不同的反应体系选择适当的指示剂。虽然以酸碱指示剂指示终点与酸碱中和反应的实际终点 pH = 7.0 并不一致，但根据计算，这种误差对求得的酸碱浓度结果影响不大。

本实验拟用已知浓度的草酸溶液标定 NaOH 溶液的浓度；再用已知浓度的硼砂溶

液标定 HCl 溶液的浓度。

【实验仪器与药品】

酸式滴定管，碱式滴定管，托盘天平，移液管，锥形瓶。

酚酞指示液，甲基红指示液，草酸（s），硼砂（s）。

【实验内容】

1. NaOH 溶液浓度的标定

准确称取草酸固体在烧杯中用纯化水溶解后，转移至 250ml 的容量瓶中，溶解草酸的烧杯和玻璃棒再用纯化水淋洗，淋洗的水也全部转移至容量瓶中，容量瓶中的水加至刻度后摇匀，计算草酸的准确浓度，该溶液为标准草酸溶液。

取洁净的移液管和锥形瓶，吸取 25ml 标准草酸溶液于锥形瓶中，加入 2 滴酚酞指示剂，摇匀待用。

取一只碱式滴定管，先用纯化水淋洗 3 次，再用需标定的 NaOH 溶液淋洗 3 次，装入 NaOH 溶液。

右手持锥形瓶不断振摇，左手挤压碱式滴定管下端玻璃球处橡皮管，滴入 NaOH 溶液，临近终点时，随着 NaOH 溶液的加入，粉红色消失较慢，因此锥形瓶要摇动均匀，直到锥形瓶中出现的粉红色半分钟不消失则滴定达到终点。注意在滴定过程中碱液可能溅到锥形瓶内壁，因此快到终点时应用洗瓶冲洗锥形瓶内壁，以减少误差。

记录下滴定中所用的 NaOH 溶液体积，再根据标准草酸溶液浓度计算出 NaOH 溶液的浓度。

2. HCl 溶液浓度的标定

准确称取硼砂晶体在烧杯中用纯化水溶解后，转移至 250ml 的容量瓶中，溶解硼砂的烧杯和玻璃棒再用纯化水淋洗，淋洗的水也全部转移至容量瓶中，容量瓶中的水加至刻度后摇匀，计算硼砂的准确浓度，该溶液为标准硼砂溶液。

取洁净的移液管和锥形瓶，吸取 25ml 标准硼砂溶液于锥形瓶中，加入 2 滴甲基红指示剂，摇匀待用。

取一只酸式滴定管，先用纯化水淋洗 3 次，再用需标定的盐酸溶液淋洗 3 次，装入盐酸溶液。用盐酸溶液滴定锥形瓶中的硼砂溶液至终点后，记录下滴定中所用的盐酸溶液体积，再根据标准硼砂溶液浓度计算出盐酸溶液的浓度。

Experiment 3　　Acid – Base Titration Exercise

Preview

1. The principle of acid-base titration.

2. The operation of acid burette and base burette.

3. Determination of the titration end-point.

Questions

1. How to discharge air bubbles in base burette?

2. Why is it important to clean up the burette, which has been rinsed with purified water, with working solution for three times?

3. Is it necessary to dry and rinse the Erlenmeyer flask with working solution?

4. Which of the following factors will affect the result in the process of standardizing the concentration of NaOH solution?

(1) There are some air bubbles in burette.

(2) The inside of the Erlenmeyer flask was not rinsed with purified water when approaching the titration end-point.

(3) There is a drop of working solution left outside the tip of the burette when the titration is finished.

Principles

The essence of acid-base neutralization reaction is:

$$H^+ + OH^- = H_2O$$

When the reaction reaches stoichiometric point, the relationship between acid content and base content is:

$$C_a V_a = C_b V_b$$

Where C_a is the concentration of acid, C_b is the concentration of base, V_a is the volume of acid, V_b is the volume of base.

So the volume ratio of acid solution and alkaline solution in neutralization reaction can be determined by titration, thus the concentration ratio of acid solution and alkaline solution could be also determined. If the concentration of a solution is known, the concentration of the other solution could be calculated.

The titration end-point can be determined by the color change of acid-base indicator, as the indicator displays different color in different pH range. Phenolphthalein appears red at pH greater than 10.0 and colorless at pH less than 8.0, and its color changes at pH 8.0—10.0. Methyl red appears yellow at pH greater than 6.2 and red at pH less than 4.4, and its color changes at pH 4.4—6.2. The appropriate indicator should be chosen according to different reaction. The titration end-point determined by indicator doesn't exactly agree with the real end-point of acid-base neutralization reaction, but the error has little influence on the result of acid and alkaline concentration after calculation.

In this experiment, the concentration of NaOH solution is titrated with the oxalic acid solution of exactly known concentration, and the concentration of HCl solution is titrated with the borax solution of exactly known concentration.

Instruments and Chemicals

acid burette, base burette, table balance, measuring pipette, Erlenmeyer flask.

phenolphthalein indicator solution, methyl red indicator solution, oxalic acid(s), borax(s).

Procedures

1. Standardizing the concentration of NaOH solution

Dissove the oxalic acid solid weighed accurately in a beaker, transferred to 250ml volumetric flask, the beaker and the glass rod used for dissolving the oxalic acid solid are rinsed with purified water, all the water transferred to the volumetric flask, add water to the volumetric flask scale and shake up, calculate the accurate concentration of the oxalic acid standard solution.

Add 25ml oxalic acid standard solution with measuring pipette to an Erlenmeyer flask, then add 2 drops of phenolphthalein indicator and shake up.

Clean up a base burette which has been rinsed with purified water for three times and with NaOH solution for three times, then fill NaOH solution into it.

Press the glass ball in rubber tube with the left hand to make the liquid drop into the Erlenmeyer flask, shake up the Erlenmeyer flask with the right hand, near the end-point, pink color will disappear slowly. If the pink color doesn't disappear in about half a minute, it means the end-point is located. During the titration, the base may splash down the upper wall of the Erlenmeyer flask, so it should be rinse down before completing the titration.

Record the volume of NaOH solution at the end-point, the concentration of NaOH solution could be calculated according to the concentration of the oxalic acid standard solution.

2. Standardizing the concentration of HCl solution

Dissolve the borax solid weighed accurately in a beaker, transferred to 250ml volumetric flask, the beaker and the glass rod used for dissolving the borax solid are rinsed with purified water, transfer all the water to the volumetric flask, add water to the volumetric flask scale and shake up, calculate the accurate concentration of the borax standard solution.

Add 25ml borax standard solution with measuring pipette to an Erlenmeyer flask, then add 2 drops of Methyl red indicator and shake up.

Clean up a acid burette which has been rinsed with purified water for three times and with HCl solution for three times, then fill it with HCl solution. Titrate the borax solution in the Erlenmeyer flask with HCl solution. Record the volume of HCl solution at the end-point, thus the concentration of HCl solution could be calculated according to the concentration of the borax standard solution.

Vocabulary

titration	滴定	indicator	指示剂
acid burette	酸式滴定管	base burette	碱式滴定管
measuring pipette	移液管	Erlenmeyer flask	锥形瓶
Phenolphthalein	酚酞	Methyl red	甲基红
oxalic acid	草酸	borax	硼砂

（编写：刘晶莹，英文核审：肖琰）

实验四 硝酸钾溶解度曲线的绘制

【预习内容】

1. 盐类溶解度与温度的关系。
2. 盐类溶解度的测定方法。
3. 台秤、温度计和量筒的使用方法。

【思考题】

1. 如何绘制硝酸钾溶解度曲线?
2. 溶解度的测定有几种方法? 各有什么优缺点?
3. 用降低温度的方法制备饱和溶液时,为什么要在降温过程中不停搅拌?

【实验原理】

在一定温度下,一定量溶剂所制成的饱和溶液中含有的溶质的量即为该溶质在该温度下的溶解度。温度对溶解度的影响很大,某物质的溶解度曲线就是表达温度对物质溶解度的影响。

测定物质在水中的溶解度有3种简单的方法。

(1) 在一定温度下,取一定重量某物质的饱和水溶液,蒸发至干,称量所余溶质的重量,就可以算出该物质在该温度下的溶解度,这种方法不适于挥发性物质溶解度的测定。

(2) 在一定温度下,取一定重量某物质的饱和溶液,用化学分析法测定该物质的含量,就可以算出该物质在该温度下的溶解度。

(3) 在较高温度下,在一定量溶剂中溶解一定重量的某物质,然后将溶液的温度缓缓下降,当溶液中刚刚开始析出晶体或出现浑浊时,立即记下温度,由此就可以推算出溶质在该温度下的溶解度。这种方法不适于测定易形成过饱和溶液的物质的溶解度。

本实验采用第三种方法,测定出硝酸钾溶液在4个不同温度下的溶解度,从而绘制出硝酸钾溶液的溶解度曲线。

【实验仪器与药品】

托盘天平,大试管,量筒,温度计,烧杯,搅拌器,电热套。

硝酸钾 (s),氯化钠 (s),醋酸钙 (饱和溶液)。

【实验内容】

1. 盐类溶解度与温度的关系

(1) 向盛有3ml纯化水的小烧杯中加入8g研细的硝酸钾固体,观察固体是否全部溶解? 如果没有全部溶解,加热至沸腾,观察固体是否全部溶解? 再冷却至室温,观察有什么变化?

(2) 向盛有3ml纯化水的试管中加入1.5g氯化钠固体,观察固体是否溶解? 如果没有溶解，加热至沸腾，观察固体是否全部溶解？然后将上清液倒入另一只试管中令其冷却，观察有何变化？

（3）向试管中加入2ml饱和醋酸钙溶液，加热后观察有何变化？

比较以上3个实验现象，说明这些盐类的溶解度与温度有什么关系。

2. 硝酸钾溶解度的测定

（1）在台秤上称取10g纯KNO_3晶体，放入一个干燥清洁的50ml大试管中，再用量筒量取10ml纯化水倒入大试管中，然后将大试管夹在铁支台柱上，并在大试管中悬挂一个普通温度计，注意使温度计的汞球全部浸入液面以下，但勿触及管底及管壁，在大试管外套上水浴加热装置（图4-1）。

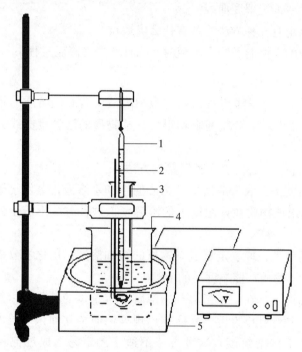

图4-1 硝酸钾溶解度装置图
1. 温度计；2. 搅拌棒；3. 大试管；4. 烧杯；5. 电热套

水浴加热的同时用搅拌器搅动KNO_3晶体，直至晶体完全溶解。将水浴撤离大试管，令大试管中的溶液自然冷却，同时应不停搅拌溶液，使溶液温度均匀下降。密切注意溶液外观的变化及温度计的温度，当溶液突然变混浊时立即记下温度。

重复操作一次，即用水浴加热，搅拌溶液，将析出的晶体重新溶解，撤去水浴，搅拌溶液自然冷却，待溶液浑浊时再记下温度。如果两次操作得到的温度相差在1℃以内，可以取平均值作为正确温度；如果两次操作得到的温度相差在1℃以上，则需要再做一次，并取三次温度的平均值作为正确温度。

（2）用量筒量取纯化水5ml，加入原来的大试管中，重复第（1）项的操作，记下温度（取两次或三次的平均值）。

（3）再加两次水，每次5ml，每次加水都要重复第（1）项的操作，记下温度（取两次或三次的平均值)。

最后一次操作完毕,将大试管中的溶液及晶体倒入指定的回收瓶中。

【实验数据的记录及处理】

1. 实验数据记录 根据称取的 KNO_3 重量和四次不同体积的水及饱和时的四个温度,就可以计算出这四个温度下 KNO_3 的溶解度。将实验记录填入表 4-1 中。

表 4-1 实验记录

(KNO_3 重量 10.0g)

实验次序	试管中水的质量	开始浑浊时溶液的温度(℃)				对应温度下 KNO_3 的溶解度(g/100g 水)
		第一次	第二次	第三次	平均值	
1						
2						
3						
4						

2. 硝酸钾溶解度曲线的绘制 根据实验中得到的四个溶解度数据,可以绘制出硝酸钾溶解度曲线。

首先选定坐标,横坐标可以选为温度,纵坐标选为溶解度,按照坐标纸的大小合理设定坐标单位,使得绘出的曲线在坐标纸的中央,美观大方。将每一个温度和它对应的溶解度在坐标纸上用实心点或空心圆等符号表示,然后绘制一条平滑的曲线,该曲线应尽可能通过或接近大多数的实验点。

【实验操作注意事项】

(1) 大试管在实验前不要洗涤,防止大试管里有水影响溶解度的测定。实验结束后要将大试管洗涤干净,倒夹于铁架台上。

(2) 安装仪器要按照由低到高的顺序进行,拆卸仪器顺序则相反。

(3) 实验过程中搅拌器要全程不停搅拌,防止局部冷热不均。

(4) 水浴所用自来水要待自来水流一会儿后再接用,防止加热后水浴变浑浊。

附录

表 4-2 不同化合物溶解度 (g/100g 水)

温度(℃)	20	30	40	50	60	70	80	100
KNO_3	31.6	45.8	63.9	85.5	110.0	138.0	169.0	242.0
NaCl	36.0	36.3	36.6	37.0	37.3	37.8	38.4	39.0
Ca(Ac)$_2$·2H$_2$O	34.7	33.8	33.2	—	32.7	—	32.5	—
Ca(Ac)$_2$·H$_2$O	—	—	—	—	—	—	—	31.1

Experiment 4　Drawing of the Solubility Curve of Potassium Nitrate

Preview

1. Relationship between salt solubility and temperature.
2. The methods of determining salt solubility.
3. The usage of table balance, measuring cylinder and thermometer.

Questions

1. How to draw the solubility curve of KNO_3?
2. How many methods are there to determine solubility? What are their advantages and disadvantages of each?
3. Why should we stir constantly in the process of cooling when lowering the temperature to prepare saturated solution?

Principles

The amount of solute in the saturated solution made with a certain amount of solvent is defined as the solubility of the solute in the temperature. Temperature has a great influence on solubility, solubility curve is the expression of temperature effecting on solubility.

There are three methods to determine the solubility of substances in water.

(1) At a certain temperature, take a certain weight of saturated solution of a substance and evaporate to dry, weigh the rest of the solute. This method is not suitable for determining the solubility of volatile substances.

(2) At a certain temperature, take a certain weight of saturated solution of a substance, determine the content of the substance using chemical analysis method.

(3) At relatively higher temperature, dissolve a certain amount of substance in a certain amount of solvent, then slowly lower the temperature of the solution. When just beginning to precipitate in the solution, immediately write down the temperature, so you can calculate the solubility of the solute at this temperature. This method is not suitable for determining the solubility of the substance easy to form supersaturated solution.

The third method is used in this experiment, we determine KNO_3 solubility at four different temperatures, and then draw the solubility curve of KNO_3.

Instruments and Chemicals

tablebalance, boiling tube, measuring cylinder, thermometer, beaker, stirrer, electric heater.

$KNO_3(s)$, $NaCl(s)$, $Ca(Ac)_2$(saturated).

Procedures

1. Relationship between salt solubility and temperature

(1) Weigh 8g KNO_3 in a small beaker, add 3ml purified water, observe whether the solid dissolved completely. If not, heat to boiling, observe whether the solid dissolved comple-

ly. Then cool to room temperature, observe what changes.

(2) Weigh 1.5g NaCl in a glass tube, add 3ml purified water, observe whether the solid dissolved completely. If not, heat to boiling, observe whether the solid dissolved completely. Then pour the supernate into the other glass tube and cool to room temperature, observe what changes.

(3) Take 2ml saturated $Ca(Ac)_2$ solution in a glass tube, heat to boiling, observe what changes.

Comparing the above three experiments, explain what the relationship between salt solubility and temperature is.

2. Determination of KNO_3 solubility

(1) Weigh 10g KNO_3 in a clean and dry boiling tube, add 10ml purified water, clip the boiling tube to a hob and hang a thermometer into it, let the location of mercury bubble below liquid level but don't touch tube bottom and wall, then heat the tube in water bath (Fig. 4-1).

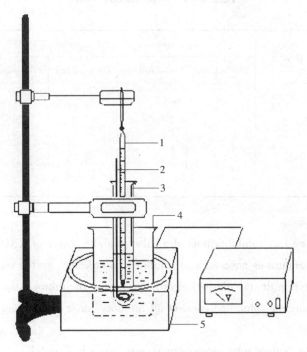

Fig. 4-1 Schematic of determining the solubility of potassium nitrate
1. thermometer; 2. stirrer; 3. boiling tube; 4. beaker; 5. electric heater

Stir and heat until all the crystals dissolves. Take the tube out of the water bath, stir the solution ceaselessly to make the solution temperature lower uniformly. Observe and write down the temperature of crystal appearing.

Repeat the operation. Heat the solution to make all the crystals dissolved and cool to crystallize, write down the temperature of crystal appearing. Take the average temperature of the operations if the temperature difference within 1℃; Repeat the operation again if the temperature difference more than 1℃, then take the average temperature of the three operations.

(2) Add 5ml purified water to the boiling tube with measuring cylinder, repeat the operation of item (1), write down the temperature (Take the average of the two or three operations).

(3) Add water twice, 5ml each time, repeat the operation of item (1), write down the temperature (take the average of the two or three operations).

Pour the solution and crystals into the designated bottle for recovered solution after the last operation.

Data record and data processing

1. Data record

Calculate the KNO_3 solubility at the four temperatures according to the weight of KNO_3 and the volume of water. Fill experimental record in table 4 − 1.

Table 4 − 1 Experimental record

(Weight of KNO_3 10.0g)

No.	The weight of water	The temperature of crystal appearing(℃)				KNO_3 solubility (g/100g H_2O)
		The first time	The second time	The third time	The average	
1						
2						
3						
4						

2. Data processing

According to the four solubility data, draw the solubility curve of KNO_3.

Take the temperature as abscissa and solubility as ordinate, set the ordinate unit reasonably to make the curve in the middle part of ordinate paper. Then draw a dot with each temperature and its corresponding solubility, connect all the dots into a smooth and clear curve.

Notes

(1) Don't wash boiling tube before experiment. Wash boiling tube clean after experiment and clip to a hob upside down.

(2) Instrument installation should be carried out in accordance with the order from down to up. Instrument remove should be carried out with the opposite order.

(3) Stir constantly during the experiment to prevent local heating.

(4) Tap water for bath should be used after tap water flows for a while, preventing that bath water becomes turbid during heating.

Appendix

Table 4-2   Salt solubility (g/100g H_2O)

Temperature(℃)	20	30	40	50	60	70	80	100
KNO_3	31.6	45.8	63.9	85.5	110.0	138.0	169.0	242.0
NaCl	36.0	36.3	36.6	37.0	37.3	37.8	38.4	39.0
$Ca(Ac)_2 \cdot 2H_2O$	34.7	33.8	33.2	—	32.7	—	32.5	—
$Ca(Ac)_2 \cdot H_2O$	—	—	—	—	—	—	—	31.1

Vocabulary

solubility	溶解度	saturated solution	饱和溶液
water bath	水浴	electric heater	电加热器
potassium nitrate	硝酸钾	calcium acetate	醋酸钙

（编写：刘晶莹，英文核审：肖琰）

第三章 化学原理实验

Chapter 3　Chemical Principle Experiments

实验五　氯化铵摩尔生成焓的测定

【预习内容】

1. 盖斯定律。
2. 测定摩尔生成焓的基本原理和方法。

【思考题】

1. 在中和焓测定过程中，为什么以 HCl 为基准进行中和焓的计算时，$NH_3 \cdot H_2O$ 必须过量？
2. 所用的量热计（包括保温杯及热电偶温度传感器等）有什么要求？是否容许有残留的洗涤水滴？为什么？
3. 试分析本实验结果产生误差主要的原因。

【实验原理】

等压反应热在数值上等于反应的摩尔焓变，因此利用杯式量热计测得等压反应热，就可以得到反应的摩尔焓变。

某温度下，由处于标准状态的各种元素的最稳定单质生成标准状态下的 1 mol 某纯物质的热效应，称为该温度下这种纯物质的标准摩尔生成焓。但是，有些物质往往不能由单质直接生成，这些物质的摩尔生成焓则无法直接测定，只能用间接的方法，可以通过测量相关反应的摩尔焓变，然后根据盖斯定律来求得。本实验就是通过测定 NH_3（aq）和 HCl（aq）反应的摩尔焓变以及 NH_4Cl（s）的摩尔溶解焓变来计算出 NH_4Cl（s）的生成焓。NH_4Cl（s）的生成可以设想通过以下途径来实现。

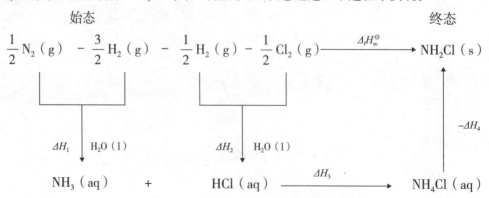

根据盖斯定律

$$\Delta H_1 + \Delta H_2 + \Delta H_3 - \Delta H_4 = \Delta H_m^\ominus \quad (5-1)$$

已知

$\Delta H_1 = -80.3 \text{kJ/mol}$（氨水在298K时的标准摩尔生成焓）

$\Delta H_2 = -167.2 \text{kJ/mol}$ [HCl（aq）在298K时的标准摩尔生成焓]

因此，只要测定ΔH_3（NH_3和HCl中和反应的摩尔焓变）及ΔH_4 [NH_4Cl（s）的摩尔溶解焓变]，利用盖斯定律即可求得NH_4Cl（s）的摩尔生成焓$\Delta_f H_m^\ominus$。

为了提高实验的准确度，减小实验误差，本实验要求$NH_3 \cdot H_2O$和HCl的中和反应在低温度溶液中进行，并在绝热、保温良好的量热计中进行。

中和反应的摩尔焓变或摩尔溶解焓变可以通过溶液的比热容和反应过程中溶液温度的改变来计算，计算公式为

$$\Delta H = -\Delta T C V d \frac{1}{n \times 1000} \quad (5-2)$$

式中，ΔT为反应前后的温度差（℃）；C为溶液的比热容 [J/(g·K)]；V为溶液的体积（ml）；d为溶液的密度（g/ml）；n为Vml溶液中NH_4Cl的物质的量。

【实验仪器与药品】

移液管（25ml×2；50ml×1），带盖的塑料保温杯（1个），热电偶数字温度计（1套，精确至0.1℃）。

HCl（1.5mol/L），NH_3（2.0mol/L），NH_4Cl（s）。

【实验内容】

1. 测量HCl（aq）和NH_3（aq）中和反应的摩尔焓变

用移液管取25.0ml的1.5mol/L HCl溶液放入预先洗净且干燥的塑料保温杯中，盖上盖子，在盖子上插入热电偶温度传感器（图5-1），水平旋转方式摇动塑料杯，直至数字温度计显示温度保持恒定为止（需要3~5分钟），记下中和反应前的温度。

再用移液管从保温杯盖子上的小孔中放入25.0ml的2.0mol/L的NH_3溶液，立即盖上小软木塞，水平旋转方式摇动保温杯，并记下中和反应后上升的最高温度（20~30秒完成）。

测定完毕后，把保温杯中的NH_4Cl溶液倒入回收瓶中，并把保温杯中的热电偶温度传感器等洗净擦干，以备下次使用。

2. 测量NH_4Cl（s）的摩尔溶解焓变

在电子天平上精确称取5.5g NH_4Cl（s）备用。用移液管量取50.0ml纯化水放入保温杯中，盖上盖子，插入热电偶温度传感器，盖上软木塞。水平旋转方式摇动保温杯，直至保温杯中的水温不再改变为止（需要3~5分钟），记下水温。

迅速将称取好的NH_4Cl（s）倒入保温杯中，立即盖紧盖子并不断以水平旋转方式轻轻地摇动保温杯，直到温度下降达到稳定的最低温度后，记下溶解后的水温。

测量完毕，把保温杯中的溶液倒入回收瓶中，并将保温杯、热电偶温度传感器等洗净擦干，放回原处。

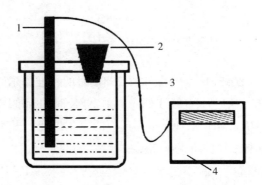

图 5-1 HCl(aq)和 NH₃(aq)中和反应焓的测定装置
1. 热电偶温度传感器；2. 软木塞；3. 塑料保温杯；4. 数字温度计

【数据记录和处理】
1. HCl 和 $NH_3 \cdot H_2O$ 的中和焓

反应物温度 T（℃）	中和反应前				中和反应后溶液的最高温度 T（℃）	中和反应的升温 ΔT（℃）
	HCl		$NH_3 \cdot H_2O$			
	浓度（mol/L）	体积（ml）	浓度（mol/L）	体积（ml）		

2. NH_4Cl 的溶解焓

无水 NH_4Cl 的摩尔质量（g/mol）	无水 NH_4Cl 的质量（g）	溶解 NH_4Cl 前的蒸馏水温度 T_1（℃）	溶解 NH_4Cl 后溶液的最低温度 T_2（℃）	$\Delta T = (T_1 - T_2)$（℃）

3. 反应焓 $\Delta H_{中和}$ 和 $\Delta H_{溶解}$ 的计算 [根据公式 (5-2)]
设：溶液的比热容为 4.18 J/(g·K)；
NH_4Cl 溶液的密度 $d = 1.00$ g/ml；
反应器的热容可以忽略不计。

4. NH_4Cl(s) 的标准摩尔生成焓的计算

ΔH_m（实测）$= \Delta H_1 + \Delta H_2 + \Delta H_3 + \Delta H_4$

$\Delta H_1 = -80.3$ kJ/mol

$\Delta H_2 = -167.2$ kJ/mol

$\Delta H_3 = _____$ kJ/mol

$\Delta H_4 = _____$ kJ/mol

$\Delta H_m = _____$ kJ/mol

5. 测量的相对误差的计算

$$相对误差 = \frac{\Delta_f H_m(实测) - \Delta_f H_m^\ominus(理论)}{\Delta_f H_m^\ominus(理论)} \times 100\%$$

[已知：$\Delta_f H_m^\ominus$（理论）$= -314.4$ kJ/mol]

Experiment 5 Determining the Molar Formation Enthalpy of Ammonium Chloride

Preview

1. Hess's Law.
2. Basic principle and method of determining the molar enthalpy of formation.

Questions

1. Why must NH_3 be excessive when HCl was used as a standard during the process of determining the enthalpy of neutralization?

2. What requirements should the calorimeter (including thermal insulation cup, thermocouple temperature sensor and so on) meet? Whether is wash water residue allowed, and why?

3. Try to analyze the main causes of the experimental error.

Principles

Isobaric heat of reaction is equal to the molar enthalpy of reaction in value. Therefore, the latter can be obtained by measuring the isobaric heat of reaction with a cup type calorimeter.

At certain temperature, the change of enthalpy from the formation of 1 mol of a compound from the most stable elementary substances, with all substances in their standard states is the standard molar enthalpy of formation of the compound. But, some compounds may be not synthesized by elementary substances, so their values of formation enthalpy are impossible to determine directly. We can calculate the formation enthalpy of a compound by determining molar enthalpy change of related reactions according to Hess's Law. In this experiment, the molar formation enthalpy of $NH_4Cl(s)$ is calculated by determining the molar reaction enthalpy change of $NH_3(aq)$ and $HCl(aq)$, and the molar enthalpy of dissolution of $NH_4Cl(s)$. We can suppose that the generation of $NH_4Cl(s)$ can be achieved by the following steps.

$$\text{primary state} \qquad\qquad\qquad\qquad\qquad\qquad\qquad \text{final state}$$

$$\tfrac{1}{2}N_2(g) - \tfrac{3}{2}H_2(g) - \tfrac{1}{2}H_2(g) - \tfrac{1}{2}Cl_2(g) \xrightarrow{\Delta_f H_m^\ominus} NH_4Cl(s)$$

$$\downarrow \Delta H_1 \; H_2O(l) \qquad\qquad \downarrow \Delta H_2 \; H_2O(l) \qquad\qquad\qquad \uparrow -\Delta H_4$$

$$NH_3(aq) \quad + \quad HCl(aq) \xrightarrow{\Delta H_3} NH_4Cl(aq)$$

According to Hess's Law

$$\Delta H_1 + \Delta H_2 + \Delta H_3 - \Delta H_4 = \Delta_f H_m^\ominus \qquad\qquad (5-1)$$

It's known that:

$\Delta H_1 = -80.3$ kJ/mol (standard molar formation enthalpy of NH_3(aq) at 298K)

$\Delta H_2 = -167.2$ kJ/mol (standard molar formation enthalpy of HCl (aq) at 298K)

So, according to the Hess's Law, the molar formation enthalpy of NH_4Cl (s) ($\Delta_f H_m$) can be calculated as long as we determine ΔH_3 (the molar enthalpy change of neutralization reaction from NH_3 (aq) and HCl (aq)) and ΔH_4 (the molar enthalpy of dissolution of NH_4Cl (s)) in this experiment.

This experiment requests that the neutralization reaction of NH_3 and HCl should be underway in adiabatic and well heat-insulating calorimeter at low temperature in order to improve the accuracy and reduce the error of the experiment.

The molar enthalpy change of neutralization reaction or the molar enthalpy of dissolution can be calculated by specific heat capacity of solution and the change of solution temperature during the experiment. The formula is:

$$\Delta H = -\Delta T C V d \frac{1}{n \times 1000} \qquad (5-2)$$

In this formula, ΔT is the temperature change before and after reaction (℃); C is the specific heat capacity of solution (J/(g·K)); V is the volume of solution (ml); d is the density of solution (g/cm); n is the molar number of NH_4Cl in V ml solution.

Instruments and Chemicals

suction pipet (25ml×2; 50ml×1), the plastic vaccum cup with lid, thermocouple digital thermometer (accurate to 0.1℃).

HCl solution (1.5mol/L), NH_3 solution (2.0mol/L), NH_4Cl (s).

Procedures

1. Determine the molar enthalpy change from neutralization reaction between HCl (aq) and NH_3 (aq).

Measure 25ml 1.5mol/L HCl solution with suction pipet, then put them into a cleaned and completely dry plastic vaccum cup. Cover a lid and insert the thermocouple temperature sensor through it into the solution (Fig. 5-1). Continuously shake the plastic cup with horizontal rotation, until the temperature keeps constant (need 3—5min). Record the temperature before neutralization reaction starting.

Then add 25.0ml 2.0mol/L NH_3 (aq) with suction pipet through the hole in the lid of the vacuum cup. Immediately cork the hole and shake the cup with horizontal rotation. Observe the temperature changing. Record the highest temperature, which can be reached after neutralization reaction (this process needs 20—30s).

Pour NH_4Cl solution in the vacuum cup into recycled bottle after the experiment. Clean and dry the thermocouple temperature sensor and others in the vaccum cup to prepare for the next use.

2. Determine the molar solution enthalpy of NH_4Cl (s).

Precisely weight 5.5g NH_4Cl (s) on the electronic balance. Then, put 50ml purified water

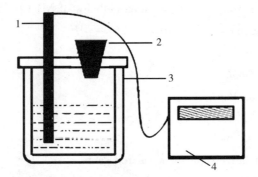

Fig. 5 – 1   Determination instrument of neutralizationenthalpy between HCl and NH$_3$
1. thermocouple temperature sensor;2. cork;3. plastic vaccum cup;4. digital thermometer

with suction pipet into the vaccum cup. Cover the lid and then insert the thermocouple temperature sensor through it into the solution. Cork the hole in the lid. Continuously shake the vaccum cup with horizontal rotation, until the temperature keeps constant (need 3—5min). Record the temperature.

Put the weighted NH$_4$Cl (s) into vaccum cup rapidly, and then tightly closed immediately. Continuously shake the plastic cup with horizontal rotation, until the temperature drops to the minimum point and keeps stable. Record the temperature after dissolution of NH$_4$Cl.

At the end of the experiment, pour the solution in the vaccum cup into recycled bottle. Clean and dry the vaccum cup, thermocouple temperature sensor and so on. Then put them back.

Data record and processing

1. The neutralization enthalpy of HCl and NH$_3$ · H$_2$O (ΔH_{neut})

Reactant temperature T(℃)	Before neutralization reaction				The maximum temperature of solution after neutralization reaction (℃)	The increasing temperature of neutralization reaction ΔT(℃)
	HCl		NH$_3$ · H$_2$O			
	Concentration (mol/L)	Volume(ml)	concentration (mol/L)	Volume(ml)		

2. The solution enthalpy of NH$_4$Cl (ΔH_{sol})

Molar mass of anhydrous NH$_4$Cl(g/mol)	Mass of anhydrous NH$_4$Cl (g)	Temperature of purified water before NH$_4$Cl dissolving T_1(℃)	The minimum temperature of solution after NH$_4$Cl dissolving T_2 (℃)	Change of temperature before and after dissolving ΔT(℃)

3. Calculation of ΔH_{neut} and ΔH_{sol} (according to formula 5 – 2)

Set: the specific heat capacity of solution is 4.18 J/(g · K);

the density of NH$_4$Cl solution: $d = 1.0$g/ml;

the heat capacity of reactor can be ignored.

4. Calculation of standard molar formation enthalpy of $NH_4Cl(s)$

$\Delta_f H_m$(measured value) $= \Delta H_1 + \Delta H_2 + \Delta H_3 - \Delta H_4$

$\Delta H_1 = -80.3$ kJ/mol

$\Delta H_2 = -167.2$ kJ/mol

$\Delta H_3 = $ _____ kJ/mol

$\Delta H_4 = $ _____ kJ/mol

$\Delta_f H_m = $ _____ kJ/mol

5. Calculation of relative error for measured value

$$\text{Relative error} = \frac{\Delta_f H_m(\text{measured value}) - \Delta_f H_m^\ominus(\text{theoretical value})}{\Delta_f H_m^\ominus(\text{theoretical value})} \times 100\%$$

($\Delta_f H_m^\ominus$(theoretical value) $= -314.4$ kJ/mol)

Vocabulary

enthalpy　　　　　焓
neutralization　　　中和
isobaric　　　　　等压的

（编写：兰阳，英文核审：夏丹丹）

实验六　化学反应速率和活化能的测定

【预习内容】

1. 基元反应、复杂反应、反应级数、质量作用定率、反应速率方程、瞬时速率及平均速率。

2. 阿仑尼乌斯公式。

【思考题】

1. 本实验测得的是平均速率还是瞬时速率？

2. 实验中加入$Na_2S_2O_3$的目的是什么？加的过多或过少，对实验结果有何影响？实验中向KI、$Na_2S_2O_3$、淀粉混合溶液中加入$(NH_4)_2S_2O_8$时为什么要快？

3. 若不用$S_2O_8^{2-}$而用I^-的浓度变化来表示反应速率，则反应速率常数是否一样？具体说明。

4. 下述情况对实验有何影响？

（1）先加$(NH_4)_2S_2O_8$溶液，然后再加KI溶液。

（2）慢慢加入$(NH_4)_2S_2O_8$溶液。

【实验原理】

在水溶液中，过二硫酸铵与碘化钾发生如下氧化还原反应：

$$(NH_4)_2S_2O_8 + 2KI = (NH_4)_2SO_4 + K_2SO_4 + I_2$$

相关离子反应方程式为：

$$S_2O_8^{2-} + 3I^- = 2SO_4^{2-} + I_3^- \tag{1}$$

此反应的速率方程为：

$$v = k \cdot c^m(S_2O_8^{2-}) \cdot c^n(I^-) \tag{6-1}$$

式中，k——反应速率常数，$m + n$——反应级数，v——瞬时反应速率；当 $c(S_2O_8^{2-})$、$c(I^-)$ 为初始浓度时，v 为起始反应速率。

测定在 Δt 时间内 $S_2O_8^{2-}$ 浓度的改变值 $\Delta c(S_2O_8^{2-})$，则平均反应速率为：

$$\bar{v} = \frac{-\Delta c(S_2O_8^{2-})}{\Delta t}$$

近似地用平均反应速率代替起始反应速率，则：

$$v = -\frac{\Delta c(S_2O_8^{2-})}{\Delta t} = k \cdot c^m(S_2O_8^{2-}) \cdot c^n(I^-) \tag{6-2}$$

为了测出 $\bar{v}$，可在将 $(NH_4)_2S_2O_8$ 溶液与 KI 溶液混合之前，在 KI 溶液中加入定量的 $Na_2S_2O_3$ 和作为指示剂的淀粉溶液。这样，在反应（1）进行的同时，也发生如下反应：

$$2S_2O_3^{2-} + I_3^- = S_4O_6^{2-} + 3I^- \tag{2}$$

反应（2）比反应（1）快得多，几乎瞬间完成，所以反应（1）生成的 I_3^- 会立即与 $S_2O_3^{2-}$ 作用，生成无色的 I^- 和 $S_4O_6^{2-}$。在 $S_2O_3^{2-}$ 没有耗尽之前，看不到碘与淀粉作用而显示的特有蓝色。而当 $Na_2S_2O_3$ 耗尽，反应（1）继续产生的微量 I_3^- 就立即与淀粉作用，使溶液呈现蓝色。

从反应方程式（1）和（2），可以看出：

$$\Delta c(S_2O_8^{2-}) = \frac{\Delta c(S_2O_3^{2-})}{2}$$

所以从所加入的 $Na_2S_2O_3$ 的量和反应出现蓝色的时间可求出反应（1）的平均速率。

将式 6-1 两边做对数，可以得到：

$$\lg v = m\lg c(S_2O_8^{2-}) + n\lg c(I^-) + \lg k$$

当 $c(I^-)$ 固定时，以 $\lg v$ 对 $\lg c(S_2O_8^{2-})$ 作图，可以得到斜率 m；当 $c(S_2O_8^{2-})$ 固定时，以 $\lg v$ 对 $\lg c(I^-)$ 作图，可以得到斜率 n。

将求得的 m 和 n，代入式 6-1 中，可求得一定温度条件下的反应速率常数 k。

反应速率常数 k 与反应温度 T 一般有如下关系，即阿仑尼乌斯公式：

$$\lg k = A - \frac{E_a}{2.303RT}$$

式中，E_a——反应的活化能，R——气体常数，T——绝对温度，A——经验常数。

通过测定不同温度时的 k 值，以 $\lg k$ 对 $\frac{1}{T}$ 作图，可得一直线，直线斜率为 $-\frac{E_a}{2.303R}$，从斜率值即可求得 E_a。

【实验仪器与药品】

烧杯（100ml），量筒（20ml×2、10ml×2），温度计（0℃~100℃），秒表，恒温水浴。

KI（0.20mol/L），$(NH_4)_2S_2O_8$（0.20mol/L），$Na_2S_2O_3$（0.010mol/L），KNO_3（0.20mol/L），$(NH_4)_2SO_4$（0.20mol/L），淀粉溶液（0.2%），$Cu(NO_3)_2$（0.20mol/L）。

【实验内容】

1. 浓度对化学反应速率的影响

在室温下，按表6－1所示的量，用贴好标签的量筒量取KI、$Na_2S_2O_3$、淀粉溶液，置于已编号的100ml烧杯中混合，搅匀，然后量取$(NH_4)_2S_2O_8$溶液，迅速加入烧杯中，同时启动秒表并不断搅拌，待溶液刚一出现蓝色，立即按停秒表，记录反应时间于表6－1中。为使每次实验的离子强度和总体积保持不变，实验2～5中减少的$(NH_4)_2S_2O_8$和KI的用量，分别用KNO_3和$(NH_4)_2SO_4$溶液补足。

2. 温度对化学反应速率的影响

按表6－1中编号4的试剂用量，将KI、$Na_2S_2O_3$、KNO_3、淀粉溶液加到100ml烧杯中，混匀，再把$(NH_4)_2S_2O_8$溶液加到另一支大试管中，然后将它们同时放入恒温水浴中，水浴温度选择在比室温高约10℃，当它们温度恒定时，记录温度，然后将$(NH_4)_2S_2O_8$溶液迅速加到KI等的混合溶液中，立即启动秒表，并不断搅拌，待出现蓝色时，按停秒表，记时。

然后再将水浴温度提高到高于室温20℃、30℃、40℃重复上述实验（编号4），记录反应时间和温度于表6－2中。

3. 催化剂对化学反应速率的影响

按表6－1中编号4中的试剂用量，将KI、$Na_2S_2O_3$、KNO_3、淀粉溶液加到100ml烧杯中，再加入2滴$Cu(NO_3)_2$溶液，然后迅速加入$(NH_4)_2S_2O_8$溶液，搅拌，立即记时，将此实验的反应速率与不加催化剂的反应速率比较，得出什么结论？

【实验数据记录与处理】

1. 数据记录

表6－1 浓度对化学反应速率的影响（室温_____℃）

试剂用量（ml）	实验编号	1	2	3	4	5
	0.20mol/L $(NH_4)_2S_2O_8$	20.0	10.0	5.0	20.0	20.0
	0.20mol/L KI	20.0	20.0	20.0	10.0	5.0
	0.010mol/L $Na_2S_2O_3$	8.0	8.0	8.0	8.0	8.0
	0.2% 淀粉	4.0	4.0	4.0	4.0	4.0
	0.20mol/L $(NH_4)_2SO_4$		10.0	15.0		
	0.20mol/L KNO_3				10.0	15.0
试剂的初始浓度（mol/L）	$Na_2S_2O_3$					
	KI					
	$(NH_4)_2S_2O_8$					
反应时间Δt（s）						
反应速率v［mol/(L·s)］						
反应速率常数						

续表

实验编号	1	2	3	4	5
平均反应速率常数					
反应级数	$m=$			$n=$	

表6-2　温度对化学反应速率的影响

实验编号	6 ($t+40$)℃	7 ($t+30$)℃	8 ($t+20$)℃	9 ($t+10$)℃
反应温度 T（K）				
反应时间 Δt（s）				
反应速率 v [mol/（L·s）]				
反应速率常数 k [L/（mol·s）]				
lg$k$				
$1/T \times 10^3$				
E_a（kJ/mol）				

2. 数据处理

（1）反应级数的计算　将表6-1中编号1和3的实验结果分别代入式6-1中，可以得到：

$$\frac{v_1}{v_3} = \frac{k \cdot c_1^m(S_2O_8^{2-}) \cdot c_1^n(I^-)}{k \cdot c_3^m(S_2O_8^{2-}) \cdot c_3^n(I^-)}，因为 c_1^n(I^-) = c_3^n(I^-)$$

所以 $\dfrac{v_1}{v_3} = \dfrac{c_1^m(S_2O_8^{2-})}{c_3^m(S_2O_8^{2-})}$

又因为 v_1、v_3 已测得，$c_1^m(S_2O_8^{2-})$、$c_3^m(S_2O_8^{2-})$ 已知，即可求出 m，同理可求 n。

（2）速率常数和活化能的计算　根据式6-1即可求得反应的速率常数 k。用 lg k 对 $\dfrac{1}{T}$ 作图，从直线的斜率求出 E_a。

Experiment 6　Determining the Rate and Activation Energy of Chemical Reaction

Preview

1. Elementary reaction, complex reaction, reaction order, Law of mass action, rate equation of reaction, momentary rate and average rate.

2. The Arrhenius equation.

Questions

1. Which rate of reaction is determined in this experiment, average or momentary?

2. What is the purpose to add $Na_2S_2O_3$ solution in the experiment? And what will happen in this experiment if the volume of $Na_2S_2O_3$ solution is varying too much? Why should we add

$(NH_4)_2S_2O_8$ solution to the mixture of KI, $Na_2S_2O_3$ and starch quickly?

3. Does the value of rate constant k still keep same if $\Delta c(I^-)$ is taken into the equation of reaction instead of $\Delta c(S_2O_8^{2-})$? Please illustrate it in detail.

4. How will conditions affect the results in the following processes?

(1) Adding $(NH_4)_2S_2O_8$ solution to a beaker followed by adding KI solution.

(2) Adding $(NH_4)_2S_2O_8$ solution slowly.

Principles

In an aqueous solution of $(NH_4)_2S_2O_8$ and KI, the following redox reaction takes place:

$$(NH_4)_2S_2O_8 + 2KI = (NH_4)_2SO_4 + K_2SO_4 + I_2$$

And the related ion equation is given as:

$$S_2O_8^{2-} + 3I^- = 2SO_4^{2-} + I_3^- \tag{1}$$

The rate of this reaction can be expressed as:

$$v = k \cdot c^m(S_2O_8^{2-}) \cdot c^n(I^-) \tag{6-1}$$

here, k is reaction rate constant; $m + n$ are reactant orders; v is momentary reaction rate. Therefore, v will be the initial reaction rate if $c(S_2O_8^{2-})$ and $c(I^-)$ are initial concentrations of $S_2O_8^{2-}$ and I^- respectively.

$\Delta c(S_2O_8^{2-})$, which is the variation of $S_2O_8^{2-}$ in a time interval Δt, will be determined in the experiment. Then the average rate of the reaction can be written as:

$$\bar{v} = \frac{-\Delta c(S_2O_8^{2-})}{\Delta t}$$

If the initial rate of the reaction is substituted by the average rate approximately, the rate equation can be put in the form:

$$v = -\frac{\Delta c(S_2O_8^{2-})}{\Delta t} = k \cdot c^m(S_2O_8^{2-}) \cdot c^n(I^-) \tag{6-2}$$

In order to determine the value of $\bar{v}$, adding starch solution as an indicator and definite amount of $Na_2S_2O_3$ into KI solution before mixing $(NH_4)_2S_2O_8$ and KI solution. So when reaction (1) is going on, the other reaction also exists:

$$2S_2O_3^{2-} + I_3^- = S_4O_6^{2-} + 3I^- \tag{2}$$

The rate of reaction (2) is much greater than that of reaction (1), which could be done almost instantaneously. So, I_3^- the product of reaction (1) will react with $S_2O_3^{2-}$ immediately to form colorless $S_4O_6^{2-}$ and I^- ions. As soon as the reactant $S_2O_3^{2-}$ is exhausted, the trace I_3^- released from reaction (1) will immediately react with starch and the color of the solution will turn to blue.

From the reaction (1) and (2), we can write:

$$\Delta c(S_2O_8^{2-}) = \frac{\Delta c(S_2O_3^{2-})}{2}$$

Thus, according to the amount of added $Na_2S_2O_3$ and the time interval from the beginning of the reaction until the color of solution turns blue, the average rate of the reaction (I) can be calculated.

Taking logarithm to both sides of equation (6-1), we have
$$\lg v = m\lg c(S_2O_8^{2-}) + n\lg c(I^-) + \lg k$$

When $c(I^-)$ is constant, plot $\lg v$ vs. $\lg c(S_2O_8^{2-})$ and we can get the value of the slope m; when $c(S_2O_8^{2-})$ is constant, plot $\lg v$ vs. $\lg c(I^-)$ and then we can get the value of the slope n.

Bring m and n into equation 6-1, we can calculate rate constant of the reaction k at a certain temperature.

With reaction rate constant k and reaction temperature T, we can predict the activation energy of chemical reaction, and then the following form (Arrhenius equation) is obtained:
$$\lg k = A - \frac{E_a}{2.303RT}$$

In this equation, E_a is activation energy of the reaction; R is universal gas constant; T is absolute temperature; A is characteristic constant.

A plot of $\lg k$ as a function of $\frac{1}{T}$ should be a straight line with the slope of $-\frac{E_a}{2.303R}$ by determining the values of k under different temperatures. So the value of E_a is obtained.

Instruments and Chemicals

beaker (100ml), measuring cylinder (20ml×2, 10ml×2), thermometer(0℃—100℃), stopwatch, water bath kettle.

KI (0.20mol/L), $(NH_4)_2S_2O_8$ (0.20mol/L), $Na_2S_2O_3$ (0.010mol/L), KNO_3 (0.20mol/L), $(NH_4)_2SO_4$ (0.20mol/L), starch solution (0.2%), $Cu(NO_3)_2$ (0.20mol/L).

Procedures

1. Concentration effect on the rate of the reaction

At room temperature, measure KI, $Na_2S_2O_3$ and starch solution with definite amount according to No. 1 in Table 6-1 with labeled measuring cylinder. Then put them into a beaker (100ml) and mix thoroughly. Then add 20ml 0.20mol/L $(NH_4)_2S_2O_8$ solution to the beaker quickly, turning on stopwatch at the same time, and stirring continuously. As soon as the color of solution turns blue, please turn off the stopwatch, and write down the passing time interval Δt in Table 6-1. Do the rest four experiments according to the given amount in Table 6-1 in the same way. In order to keep the ionic strength and the total volume constant in each portion of the mixture, KNO_3 and $(NH_4)_2SO_4$ solution are added to supplement the inadequate part of $(NH_4)_2S_2O_8$ and KI, respectively.

2. Temperature effect on the rate of the reaction

Mix accurately weighed KI, $Na_2S_2O_3$, KNO_3 and starch solution according to No. 4 in Table 6-1 thoroughly, which is contained in a beaker (100ml). Then add $(NH_4)_2SO_4$ solution into a big tube. Put the beaker and tube into a water bath simultaneously at the temperature 10℃ higher than room temperature. When the temperature is constant, record the temperature. Then add $(NH_4)_2S_2O_8$ solution in the mixture containing KI and so on, and turn on the stopwatch, stirring continuously. Record the passing time interval Δt when the solution just turns blue.

At the temperature 20℃, 30℃, 40℃ higher than room temperature, repeat the above experiment (No. 4). Record the passing time interval and temperature, and fill in Table 6-2.

3. Catalyst effect on the rate of the reaction

Add definite amount of KI, $Na_2S_2O_3$, KNO_3 and starch solution to a beaker (100ml) according to No. 4 in Table 6-1. Subsequently, add two drops of $Cu(NO_3)_2$ solutions to the above mixture. And then $(NH_4)_2S_2O_8$ was added rapidly. Stir the solution, and record the passing time interval. What conclusion can be obtained through comparing the reaction rate in this experiment with that in the experiment without catalyst.

Data record and processing

1. Data record

Table 6-1 Concentration effect on the chemical reaction rate (room temperature _____ ℃)

Experiment number		1	2	3	4	5
Reagents (ml)	0.20mol/L $(NH_4)_2S_2O_8$	20.0	10.0	5.0	20.0	20.0
	0.20mol/L KI	20.0	20.0	20.0	10.0	5.0
	0.010mol/L $Na_2S_2O_3$	8.0	8.0	8.0	8.0	8.0
	0.2% starch solution	4.0	4.0	4.0	4.0	4.0
	0.20mol/L $(NH_4)_2SO_4$		10.0	15.0		
	0.20mol/L KNO_3				10.0	15.0
Initial concentration of reagents (mol/L)	$Na_2S_2O_3$					
	KI					
	$(NH_4)_2S_2O_8$					
Δt(s)						
V(mol/(L·s))						
k						
$\bar{k}$						
Reactant order		$m=$			$n=$	

Table 6-2 Temperature effect on the reaction rate

Experiment number	6($t+40$)℃	7($t+30$)℃	8($t+20$)℃	9($t+10$)℃
T(K)				
Δt(s)				
V(mol/(L·s))				
k				
lg k				
$1/T$				
E_a				

2. Data processing

(1) Calculation of the reactant order

Putting the results of No. 1 and No. 3 in Table 6-1 into equation 6-1, then the following form is obtained:

$$\frac{v_1}{v_3} = \frac{k \cdot c_1^m(S_2O_8^{2-}) \cdot c_1^n(I^-)}{k \cdot c_3^m(S_2O_8^{2-}) \cdot c_3^n(I^-)}$$

$\because c_1^n(I^-) = c_3^n(I^-)$

$\therefore \dfrac{v_1}{v_3} = \dfrac{c_1^m(S_2O_8^{2-})}{c_3^m(S_2O_8^{2-})}$

Moreover, m can be calculated as that v_1 and v_3 has been determined, $c_1^m(S_2O_8^{2-})$ and $c_3^m(S_2O_8^{2-})$ are known.

(2) Calculation of the rate constant and activation energy

The reaction rate constant k can be calculated by equation 6-1. Plot lg k vs. $1/T$ on the graph paper, and the proportion and configuration should be apposite. The value of E_a can be calculated by the slope.

Vocabulary

constant	常数
momentary	瞬时的
starch	淀粉
indicator	指示剂

（编写：兰阳，英文核审：夏丹丹）

实验七 凝固点降低法测定葡萄糖的摩尔质量

【预习内容】

1. 理解稀溶液的依数性。
2. 掌握凝固点降低法测定溶质摩尔质量的原理和方法。

【思考题】

1. 稀溶液有哪些性质，它们是怎样产生的，有何共性？
2. 冰盐水为什么能做冷冻剂？为什么不能单独用冰？为什么加盐？
3. 凝固点降低法测定葡萄糖摩尔质量，为什么必须测定水的凝固点？

【实验原理】

稀溶液依数性是指稀溶液的性质和在一定量的溶剂中所溶解的溶质分子数呈正比而和溶质的本性无关，主要包括：溶液的凝固点下降、沸点上升、蒸气压下降和渗透压。

稀溶液的凝固点决定于其浓度和溶剂的性质。凝固点降低与溶液的质量摩尔浓度（b_B）呈正比：

$$\Delta T_f = T_f^* - T_f = K_f \cdot b_B \quad (7-1)$$

式中，T_f^* 是纯溶剂的凝固点，T_f 是溶液的凝固点。K_f 称为凝固点降低常数。K_f 只取决于溶剂的性质，与溶质的性质无关。

可以通过测定溶液凝固点的降低来测定溶质的摩尔质量 M_B。

$$b_B = \frac{m_B/M_B}{m_A} \quad (7-2)$$

将式（7-2）带入式（7-1）可得：

$$M_B = \frac{K_f m_B}{m_A \Delta T_f} \quad (7-3)$$

【仪器与药品】

温度计（1/10 温度计的刻度可读至 0.01℃，准确到 0.1℃），移液管（25ml），电子天平，试管（盛放待测液），空气套管，细搅拌棒（上下搅动测定液，使温度均匀），橡皮塞，粗搅拌棒（搅拌冰水），厚壁烧杯（500ml，内盛冰盐水以降低温度）。

蒸馏水，葡萄糖，粗盐，冰。

【实验步骤】

1. 测量葡萄糖溶液的凝固点 T_f

（1）将适量冰块和水放入烧杯（两者占烧杯容积的 3/4），然后加入一定量粗盐，使冰盐水混合物的温度达 -5℃以下。在实验中随时补充冰块，取出多余的水，并上下搅拌冰块，以保持温度恒定。

（2）用分析天平称取 2.3~2.5g（精确到 ±0.0001g）葡萄糖。

（3）将葡萄糖全部转移到干燥的凝固点测定管中，顺着管内壁用移液管加入 25ml 蒸馏水，轻轻摇匀（注意不要将溶液溅出）。待葡萄糖完全溶解后，盖上带有温度计和细搅拌棒的橡皮塞，温度计的水银球应全部浸入溶液。将测定管插入空气套管内，放入大烧杯中（图7-1）。用细搅拌棒慢慢搅拌（每秒一次）葡萄糖溶液，搅拌棒尽量避免接触试管内壁和温度计，否则因摩擦产生的热量将有可能影响测定结果。待试管中出现冰屑，温度回升，记录在试管中温度计回升后的最高温度。取出试管，流水冲洗试管外壁，使冰屑完全融化，重复上述操作一次。直到前后两次温差不超过 0.02℃，即得葡萄糖溶液的凝固点 T_f。

2. 测定纯溶剂（水）的凝固点 T_f^* 弃去试管内溶液，用自来水洗净，并用蒸馏水洗涤 3 次，用量筒量取约 25ml 蒸馏水放入试管中，按照步骤 1 的方法测定纯水的凝固点 T_f^*。

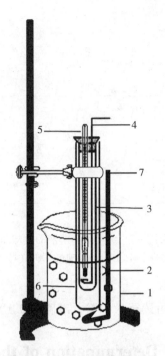

图 7-1　冰点测定装置图

1. 烧杯；2. 冰块；3. 测量管；4. 细搅拌棒；5. 温度计；6. 空气套管；7. 粗搅拌棒

【数据记录与结果处理】

表 7-1　实验记录

测定次数	凝固点（℃）		溶质质量（g）	溶剂质量（g）	ΔT_f
	蒸馏水	葡萄糖溶液			
1					
2					
3					
结果计算：M =					

水的 ρ 和 K_f 值分别为：$\rho = 1.000 \text{g/ml}$，$K_f = 1.86 \text{K·kg/mol}$

【注意事项】

1. 凝固点　凝固点的温度并不像我们想象的那样容易读出，因为溶液的温度往往冷却到冰点时还不结冰，只有使其温度继续下降到某一温度时才开始析出冰来，同时因放出大量的熔化热而使溶液的温度突然上升，上升的最高点就是冰点（图 7-2）。

2. 冰盐冷冻剂　当食盐、冰及少量水混合在一起，因为在同温下冰的蒸气压大于饱和食盐水的蒸气压，故冰要熔化，熔化时就要吸收周围的热量，故冰盐水可做冷冻剂，最低可降至 -22℃，若为"$CaCl_2$ 冰水"冷冻剂，最低可达 -55℃。

3. 使用 1/10 刻度温度计必须注意

（1）这种温度计很贵重而且很长，使用和放置均要注意，避免碰断。

（2）这种温度计水银球处玻璃薄，不能用力捏，若被冻住，不可用力拔，应用手

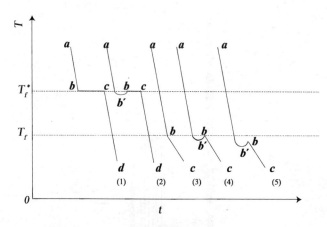

图 7-2 纯溶剂和溶液的冷却曲线

(1) 溶剂理想冷却；(2) 溶剂实验冷却；(3) 溶液理想冷却；(4) 溶液实验冷却；(5) 溶液严重过冷

握试管，待冰熔化后才能取出。

(3) 1/10 温度计的刻度可读至 0.01℃，准确到 0.1℃。

Experiment 7 Determination of the Molar Mass of Glucose by Freezing Point Depression Method

Preview

1. To understand the colligative properties of diluted solution.

2. To grasp the principle and methods for determining the molar mass of solute by freezing point depression.

Questions

1. Please outline the properties of diluted solution, and consider the basic principles of these properties. Are there any common points of these properties?

2. Why brine ice (ice-salt mixture), instead of only ice, can be used as the refrigerant?

3. Why is it necessary to determine the freezing point of water during the process that gets the molar mass of glucose in this work?

Principles

Colligative properties are those properties of solutions that depend on the number of dissolved particles in solution, but not on the identities of the solutes. The four commonly studied colligative properties are freezing point depression, boiling point elevation, vapor pressure lowering, and osmotic pressure.

The freezing point is associated with solvent and the concentration of solution. The freezing point depression ΔT_f is found to be proportional to the molality (b_B) of the solution:

$$\Delta T_f = T_f^* - T_f = K_f \cdot b_B \tag{7-1}$$

Where, K_f is cryoscopic constant, which is only determined by the property of the solvent, and has nothing to do with the property of the solute.

The freezing point of solution (T_f) is lower than that of its pure solvent (T_f^*). For a non-electrolytic dilute solution, we have the following equations:

$$b_B = \frac{m_B/M_B}{m_A} \qquad (7-2)$$

According to this, the molar mass of solute can be given as:

$$M_B = \frac{K_f m_B}{m_A \Delta T_f} \qquad (7-3)$$

Instruments and Chemicals

1/10 scale thermometer (read to the second decimal digit), volumetric pipette (25ml), analytical balance, test tube (fill with the solution to be determined), air coated tube (determine the freezing point accurately), thin stirring rod, rubber stopper, thick stirring rod (stir ice-salt water), thick wall beaker (500ml, fill with ice-salt water to decrease the temperature).

distilled water, glucose, crude salt, ice.

Procedure

1. Determination the freezing point of the glucose solution.

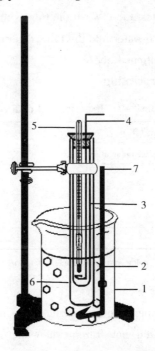

Fig. 7-1 Schematic of freezing point determination
1. breaker; 2. ice; 3. determing tube; 4. thin stirring rod;
5. thermometer; 6. air coated tube; 7. thick stiring rod

(1) Fill some pieces of ice and a small quantity of water into a thick walled beaker to about 3/4 full. Add certain amount of crude salts to decrease the temperature. Keep the temperature of this ice-salt-water mixture below $-5°C$.

(2) Accurately weigh 2.3—2.5g (precision ±0.0001g) glucose with an analytical balance.

(3) Put the glucose into a dry test tube, and add 25ml distilled water along the inner wall of the tube with a volumetric pipette, shaking slightly. Be careful not to splash the solution out. Plug the tube by a rubber stopper with a thermometer and a thin stirring rod after the glucose is dissolved completely, and then put the built-up test tube into an air coated tube and place the whole device into the thick wall beaker (Fig. 7-1), stir the glucose solution with thin stirring rod slightly. Add ice and withdraw water from the beaker if necessary. To reduce any heating effect from friction on the results, the contact of the stirring rod to the wall of the tube and thermometer during the stirring process should be carefully avoided. Stir the glucose solution continuously until some pieces of ice appear. Continue to stir and observe the thermometer carefully as the temperature is rising. When the temperature no longer raises, record the temperature (freezing point). Take out the test tube and rinse the outside of the test tube with tap water until the ice in it melts. Repeat the above procedures until the difference of two determinations is not exceed 0.02℃, take the average value (T_f).

2. Determine the freezing point of the pure solvent (water).

Discard the solution in the test tube. Wash the tube with tap water and rinse it 3 times with distilled water. Add 25ml distilled water into the tube, determine the freezing point (T_f^*) of the water with the same method mentioned above.

Data record and result processing

Table 7-1   Experimental data records

Experiment number	Freezing point (℃)		solute(g)	solvent(g)	ΔT_f
	Distilled water	Glucose solution			
1					
2					
3					

Result: M =

The values of ρ and K_f of water are as below: $\rho = 1.000$ g/ml, $K_f = 1.86$ K·kg/mol

Notes

1. Freezing Point: The temperature of freezing point is not as easy readout as we expected, because the solution is often not frozen when the temperature is cooled to freezing point. Continued decline to certain temperature is necessary when the solution begins to freeze, and the emission of a lot of melting heat will lead to a sudden rise on temperature, the highest point is the freezing point (Fig. 7-2).

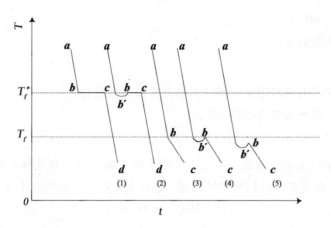

Fig. 7-2 The cooling curve of puresolvent and solution
1. solvent ideal cooling; 2. solvent practical cooling; 3. solution ideal cooling;
4. solution practical cooling; 5. solution exorbitant super cooling

2. Ice-salt freezing refrigerants: When sodium chloride, ice, and a small amount of water were mixed together, as the vapor pressure of ice is greater than that of saturated sodium chloride aqueous solution at the same temperature, the ice will melt and absorb heat from the surrounding, the ice-salt can work as refrigerants with the minimum temperature to -22℃. For "$CaCl_2$ ice" refrigerants, the minimum temperature of -55℃ can be reached.

3. Pay attention when 1/10 scale thermometer is used:

(1) This thermometer is very precious and long, care should be taken for its use and placement to avoid breaking.

(2) The wall of the mercury bulb is thin and fragile. If it is frozen in ice, don't pull out with force. Holding the test tube with hand, until the ice melt and remove it.

(3) The accuracy of 1/10 scale is 0.1℃, a value of 0.01℃ can be read.

Vocabulary

glucose	葡萄糖	molar mass	摩尔质量
freezing point	凝固点	colligative properties	依数性
diluted solution	稀溶液	thermometer	温度计
air coated tube	空气套管	analytical balance	分析天平
stirring rod	搅拌棒	rubber stopper	橡皮塞

（编写：王绍宁，英文核审：夏丹丹）

实验八 电位法测定硼酸的电离常数

【预习内容】

1. 掌握用电位滴定测定电离常数的方法。

2. 学习使用 pH 计。
3. 练习移液管的使用。

【思考题】

1. 如果改变所测硼酸溶液的温度，电离常数有无变化？
2. 如何测定弱碱的电离指数 pK_b？

【实验原理】

根据 Bronsted – Lowry 理论，酸是质子给予体，碱是质子接受体。弱酸的电离实质上是一个质子传递反应，也称作酸碱反应。弱酸 HA 在水中达到如下平衡：

$$HA + H_2O = H_3O^+ + A^-$$

简写成：

$$HA = H^+ + A^-$$

在一定温度下，当溶液中建立平衡时，存在下面的关系：

$$K_a = \frac{[H^+][A^-]}{[HA]} \tag{8-1}$$

两边取负对数得：

$$pK_a = pH + \lg\frac{[HA]}{[A^-]} \tag{8-2}$$

向弱酸 HA 溶液中加入金属氢氧化物（MeOH，如氢氧化钾、氢氧化钠等）进行中和反应，MeOH 作为强电解质在水溶液中可以完全解离成 Me$^+$ 和 OH$^-$。OH$^-$ 作为强碱与 HA 反应得到如下反应式：

$$HA + OH^- = A^- + H_2O$$

根据物料平衡我们可以得到：

$$[HA] = c(HA) - c(MeOH) \tag{8-3}$$

$$[A^-] = c(MeOH) \tag{8-4}$$

则 $\frac{[HA]}{[A^-]}$ 可以写成：

$$\frac{[HA]}{[A^-]} = \frac{c(HA) - c(MeOH)}{c(MeOH)} \tag{8-5}$$

由此，当加入不同量的碱液时部分 HA 被中和，从中和的程度可知溶液中 $\frac{[HA]}{[A^-]}$ 的比值。通过测定溶液的 pH 值，并根据公式 8-2 即可计算出该弱酸的 pK_a 值。

【仪器与药品】

pHS – 25 型酸度计，磁力搅拌器，移液管（100ml），吸量管（1ml），烧杯（250ml），H_3BO_3（0.01mol/L），NaOH（0.100mol/L）。

【实验步骤】

用移液管精密量取 0.01mol/L 的 H_3BO_3 溶液 100ml，置于洁净、干燥的 250ml 烧杯中，然后用吸量管，逐次精密量取 0.100mol/L NaOH 1.0ml，分别加入到 H_3BO_3 溶液中，充分搅拌，并依次测出各自对应的 pH 值。

【实验数据记录及处理】

表 8－1　实验数据

0.100mol/L NaOH 体积（ml）	pH	$[HA]$ (mol/L)	$[A^-]$ (mol/L)	$\dfrac{[HA]}{[A^-]}$	$\lg\dfrac{[HA]}{[A^-]}$	$pK_a = pH + \lg\dfrac{[HA]}{[A^-]}$
1.0						
2.0						
3.0						
4.0						
5.0						
6.0						
7.0						
8.0						
9.0						

Experiment 8　Determination of Dissociation Constant of a Weak Acid by Potentiometric Method

Preview

1. To know how to determine the dissociation constant of a weak acid by potentiometric method.

2. To learn how to determine the pH values of solutions by pH meter.

3. To practice how to use the pipette correctly.

Questions

1. Does the dissociation constant change if the temperature of H_3BO_3 solution has obvious varied?

2. How to determine the dissociation index, pK_b, of a weak base?

Principle

The Bronsted-Lowry classification defines an acid as a proton donor and a base as a proton acceptor. The dissociation of a weak acid (HA) represents a proton-transfer reaction, which is also called as the acid-base reaction. Therefore the weak acid, HA, generates the following equilibrium in water:

$$HA + H_2O = H_3O^+ + A^-$$

Or

$$HA = H^+ + A^-$$

At certain temperature, the dissociation constant, K_a, is given as:

$$K_a = \frac{[H^+][A^-]}{[HA]} \tag{8-1}$$

Taking the negative logarithm of both sides of equation 8-1 yeilds:

$$pK_a = pH + \lg\frac{[HA]}{[A^-]} \tag{8-2}$$

Consider the neutralization reaction of week acid HA by metal hydroxide, MeOH (e.g. KOH, NaOH, etc). As a strong electrolyte, MeOH completely dissociate into Me^+ and OH^- ions in diluted aqueous solutions. As a strong base, the OH^- anions will react with the weak acid, HA:

$$HA + OH^- = A^- + H_2O$$

Consequently, we can obtain the following mass balance equations

$$[HA] = c(HA) - c(MeOH) \tag{8-3}$$

$$[A^-] = c(MeOH) \tag{8-4}$$

Thus,

$$\frac{[HA]}{[A^-]} = \frac{c(HA) - c(MeOH)}{c(MeOH)} \tag{8-5}$$

Accordingly, when different amounts of MeOH are added, the ratios of $\frac{[HA]}{[A^-]}$ can be derived from the degrees of neutralization of HA. Thus, when the pH of the HA solution is known, the value of pK_a for HA can be calculated via equation 8-2.

Instuments and Chemicals

pHS-25 pH meter; magnetic stirrer; volumetric pipette (100ml); mohr pipette (1ml); beaker (250ml); H_3BO_3 (0.01mol/L); NaOH (0.100mol/L).

Procedure

Transfer 100ml of 0.01mol/L H_3BO_3 solution by a volumetric pipette to a 250ml beaker, determine the pH value by a pH meter. Successively add 0.100mol/L standard NaOH solution 1.0ml by a mohr pipette, stir well, and record the pH values at each point.

Data record and result processing

Table 8-1 Experimental data records

Volume of 0.100mol/L NaOH(ml)	pH	[HA] (mol/L)	[A$^-$] (mol/L)	$\frac{[HA]}{[A^-]}$	$\lg\frac{[HA]}{[A^-]}$	$pK_a = pH + \lg\frac{[HA]}{[A^-]}$
1.0						
2.0						
3.0						
4.0						
5.0						
6.0						
7.0						
8.0						
9.0						

Vocabulary

dissociation constant	电离常数	potentiometric method	电位法
proton donor	质子给予体	proton acceptor	质子接受体
proton-transfer reaction	质子传递反应	neutralization reaction	中和反应

（编写：王绍宁，英文核审：夏丹丹）

实验九　电动势法测定氯化银的溶度积常数

【预习内容】

1. 电动势法测定难溶电解质溶度积的原理和方法。
2. 电子天平及酸度计的使用方法。

【思考题】

1. 试写出 $\varphi^{\ominus}$（AgX/Ag）与 $K_{sp}^{\ominus}$（AgX）之间的关系式。
2. 有一原电池（−）Ag｜AgX｜X⁻（c）‖ KCl（饱和）｜Hg_2Cl_2｜Hg（+），写出其 E 的表达式。
3. 如何活化银电极？

【实验原理】

利用电动势法可以测定难溶电解质的溶度积常数，如测定某卤化银溶度积常数时，首先选取两个电极和相应的溶液构成原电池，然后测定该原电池的电势差 E 值，通过 E 与 $K_{sp}^{\ominus}$ 之间的关系，即可求出该卤化银溶度积常数。

原电池：（−）Ag｜AgX｜X⁻（c）‖ KCl（饱和）｜Hg_2Cl_2｜Hg（+）

负极的电极反应为：　　AgX(s) + e $\rightleftharpoons$ Ag(s) + X⁻

根据能斯特方程式得：

$$\varphi(AgX/Ag) = \varphi^{\ominus}(AgX/Ag) - \frac{RT}{nF}\ln[c(X^-)/c^{\ominus}]$$

当 $T = 298.15K$ 时，

$$\varphi(AgX/Ag) = \varphi^{\ominus}(AgX/Ag) - \frac{0.0592}{n}\lg c(X^-)$$

其中，$\varphi^{\ominus}(AgX/Ag) = \varphi^{\ominus}(Ag^+/Ag) + 0.0592 \lg K_{sp}^{\ominus}(AgX)$

又因为，$E = \varphi(甘汞) - \varphi(AgX/Ag)$

所以，$E = \varphi(甘汞) - \varphi^{\ominus}(Ag^+/Ag) - 0.0592 \lg K_{sp}^{\ominus}(AgX) + 0.0592 \lg c(X^-)$

式中，$\varphi(甘汞)$ 和 $\varphi^{\ominus}(Ag^+/Ag)$ 是已知的，在一定温度下，$K_{sp}^{\ominus}(AgX)$ 又是一常数，所以 E 与 $\lg c(X^-)$ 呈线性关系，因此可以改变原电池体系中的 $c(X^-)$，测得相应的 E，以 E 为纵坐标，以 $\lg c(X^-)$ 为横坐标作图，得一条直线，其截距为 $\varphi(甘汞) - \varphi^{\ominus}(Ag^+/Ag) - 0.0592 \lg K_{sp}^{\ominus}(AgX)$，进而求得 $K_{sp}^{\ominus}(AgX)$。

【实验仪器与药品】

pHS-25 酸度计，饱和甘汞电极，银电极，电子天平，吸量管（1ml），移液管（50ml），烧杯（100ml），容量瓶（50ml）。

KCl（s），$AgNO_3$（0.1mol/L），HNO_3（6mol/L）。

【实验内容】

1. 溶液的配制 用 50ml 容量瓶配制 0.2000mol/L 的 KCl 溶液。

2. 电极的活化 将银电极插入 6mol/L HNO_3 溶液中，当银电极表面有气泡产生且呈银白色时，将银电极取出，先用自来水冲洗，再用纯化水洗净，滤纸吸干备用。

3. 原电池电动势的测定

（1）将银电极和饱和甘汞电极安装在电极架上，银电极接负极，甘汞电极接正极。

（2）在 100ml 干燥烧杯中，用移液管加入 50ml 纯化水，再用吸量管移入 1ml 0.2000mol/L KCl 溶液，滴入 1 滴 0.1mol/L $AgNO_3$ 溶液，摇匀后，将电极放入该溶液中，测定其电动势值 E_1，并记录于表 9-1 中。

（3）再用吸量管移取 1ml 0.2000mol/L KCl 溶液于同一烧杯中，摇匀后，测定其电动势 E_2，并记录于表 9-1 中。

（4）如此重复步骤（3），分别测得 E_3、E_4、E_5，记录于表 9-1 中。

【实验数据的记录及处理】

1. 数据记录

表 9-1 数据记录

实验编号	1	2	3	4	5
加入 KCl 溶液累计体积（ml）	1.00	2.00	3.00	4.00	5.00
$c(Cl^-)$（mol/L）					
$\lg c(Cl^-)$					
E（V）					

2. 数据处理 用 E 对 $\lg c(Cl^-)$ 作图，从直线在纵坐标上的截距求出 $K_{sp}^{\ominus}$（AgCl）。

Experiment 9 Determining Solubility Product Constant of Silver Chloride by Cell Potential Method

Preview

1. The principle and method of determining solubility product constant of insoluble electrolytes by cell potential method.

2. The usage of electronic balance and acidometer.

Questions

1. Please write out the related expression between $\varphi^{\ominus}(AgX/Ag)$ and $K_{sp}^{\ominus}(AgX)$.

2. Please write out the expression of E of the galvanic cell:

$(-)$ Ag | AgX | X$^-$ (c) || KCl (saturated) | Hg$_2$Cl$_2$ | Hg $(+)$

3. How to activate the silver electrode?

Principles

Cell potential method can be used to determine the solubility product constant of insoluble electrolytes. Taking the determination of solubility product constant of a silver halide as example, firstly, two electrodes and the corresponding solution are selected, secondly, the cell potential E is determined, finally, the solubility product constant of silver halide can be obtained through the related expression between the E and $K_{sp}^\ominus$. For example,

$(-)$ Ag | AgX | X$^-$ (c) || KCl (saturated) | Hg$_2$Cl$_2$ | Hg $(+)$

The reaction aquation of the negative electrode is shown as follows:

$$AgX(s) + e \rightleftharpoons Ag(s) + X^-$$

According to the Nernst equation:

$$\varphi(AgX/Ag) = \varphi^\ominus(AgX/Ag) - \frac{RT}{nF}\ln(c(X^-)/c^\ominus)$$

$T = 298.15K, \varphi(AgX/Ag) = \varphi^\ominus(AgX/Ag) - \frac{0.0592}{n}\ln c(X^-)$

Here, $\varphi^\ominus(AgX/Ag) = \varphi^\ominus(Ag^+/Ag) + 0.0592 \lg K_{sp}^\ominus(AgX)$

$E = \varphi(\text{calomel}) - \varphi(AgX/Ag)$

So, $E = \varphi(\text{calomel}) - \varphi^\ominus(Ag^+/Ag) - 0.0592 \lg K_{sp}^\ominus(AgX) + 0.0592 \lg c(X^-)$

Since, $\varphi(\text{calomel})$ and $\varphi^\ominus(Ag^+/Ag)$ are known, $K_{sp}^\ominus(AgX)$ is a constant at a certain temperature, there is linear relationship between E and $\lg c(X^-)$. A plot of E as a function of $\lg c(X^-)$ should be a straight line with the intercept ($\varphi(\text{calomel}) - \varphi^\ominus(Ag^+/Ag) - 0.0592 \lg K_{sp}(AgX)$) by determining E values wnder diffenent $c(X^-)$, so $K_{sp}^\ominus(AgX)$ is obtained.

Instruments and Chemicals

pHS-25 acidometer, saturated calomel electrode, silver electrode, electronic balance, mohr pipet (1ml), volumetric pipet (50ml), beaker (100ml), volumetric flask (50ml).

KCl (s), AgNO$_3$ (0.1mol/L), HNO$_3$ (6mol/L).

Procedures

1. Preparation of the solution

Prepare 0.2000mol/L solution of KCl with 50ml volumetric flask.

2. Activation of the electrode

The silver electrode was inserted into 6mol/L HNO$_3$ until air bubbles is generated on the silver electrode surface and the silver electrode becomes silverly white, then the silver electrode is washed with tap water and purified water. Then dried with filter paper in turn.

3. The determination of a galvanic cell potential

(1) Mount the silver electrode and saturated calomel electrode on the frame. Silver electrode is connected with the negative pole, calomel electrode is connected with the positive elec-

trode.

(2) Add 50ml of purified water with volumetric pipet, 1ml of 0.2000mol/L KCl solution with mohr pipet and 1 drop of 0.1mol/L $AgNO_3$ solution into a dry 100ml beaker accurately, shake well, put the electrode into the electrolytes solution, measure cell potential E_1, and fill it in table 9-1.

(3) Then pipette 1.00ml 0.2000mol/L KCl solution in the same beaker, shake well, measure its cell potential E_2 and fill it in table 9-1.

(4) Repeat step (3), E_3, E_4 and E_5 are measured respectively, fill then in table 9-1.

Data record and Data processing

1. Data record

表 9-1 Data record

No.	1	2	3	4	5
cumulative volume of KCl solution (ml)	1.00	2.00	3.00	4.00	5.00
$c(Cl^-)$ (mol/L)					
$\lg c(Cl^-)$					
$E(V)$					

2. Data processing

$K_{sp}^{\ominus}(AgX)$ can be obtained from the intercept of the straight line on E to $\lg c(Cl^-)$.

Vocabulary

solubility product constant	溶度积常数	insoluble electrolyte	难溶电解质
electric potential method.	电动势法	acido meter	型酸度计
galvanic cell	原电池	Nernst equation	能斯特方程
cell potential	电动势	silver electrode	银电极
saturated calomel electrode	饱和甘汞电极	volumetric pipet	移液管
volumetric flask	容量瓶	mohr pipet	吸量管

（编写：王鸿钢，英文核审：肖琰）

实验十 酸碱平衡和沉淀溶解平衡

【预习内容】

1. 酸碱平衡及影响平衡移动的因素。
2. 缓冲溶液的性质。
3. 沉淀的生成、溶解条件。
4. pHS-25 型酸度计的使用方法。
5. pH 试纸的使用、液体的取用。

【思考题】

1. 同离子效应与缓冲溶液的原理有何异同？
2. 如何抑制或促进水解？举例说明。
3. 是否一定要在碱性条件下，才能生成氢氧化物沉淀？不同浓度的金属离子溶液，开始生成氢氧化物沉淀时，溶液的pH值是否相同？
4. 请计算下列反应的平衡常数

(1) $Mg(OH)_2 + 2NH_4^+ \rightleftharpoons Mg^{2+} + 2NH_3 \cdot H_2O$

(2) $Fe(OH)_3 + 3NH_4^+ \rightleftharpoons Fe^{3+} + 3NH_3 \cdot H_2O$

(3) $Ag_2CrO_4 + 2Cl^- \rightleftharpoons 2AgCl + CrO_4^{2-}$

(4) $2AgCl + CrO_4^{2-} \rightleftharpoons Ag_2CrO_4 + 2Cl^-$

【实验原理】

1. 弱电解质在溶液中的电离平衡及其移动。

一元弱酸HA，在水溶液中存在下列电离平衡：

$$HA \rightleftharpoons H^+ + A^-$$

酸度常数：$K_a = \dfrac{[H^+][A^-]}{[HA]}$

其共轭碱一元弱碱A^-，在水溶液中存在下列电离平衡：

$$A^- + H_2O \rightleftharpoons HA + OH^-$$

碱度常数：$K_b = \dfrac{[HA][OH^-]}{[A^-]}$

在HA溶液中，若加入含有相同离子的强电解质，即增加A^-或者H^+离子的浓度，则平衡向生成HA分子的方向移动，使HA的电离度降低。这种作用称作同离子效应。

2. 缓冲溶液一般是由共轭酸碱对组成的。其pH值可以用下式表示：

$$pH = pK_a + \lg(c_b/c_a) \tag{1}$$

考虑活度后，其计算公式为：

$$pH = pK_a + \lg\dfrac{f_2 c_b}{f_1 c_a} \tag{2}$$

式中f_1、f_2分别为酸、碱的活度系数。若不考虑活度，计算值与实验测定值之间将产生误差。

缓冲溶液的pH值除主要决定于pK_a外，还与碱和酸的浓度比值有关。若配制缓冲溶液所用的碱和酸溶液的原始浓度相同，则配制时所取碱和酸溶液体积(V)的比值就等于它们浓度的比值。所以(1)式可改写为：

$$pH = pK_a + \lg(V_b/V_a)$$

这时只要按碱和酸溶液体积的不同比值配制溶液，就可以得到所需pH值的缓冲溶液。

稀释缓冲溶液时，溶液中的碱和酸浓度都是以相等比例降低，碱和酸浓度比值不改变，因此适当稀释不影响缓冲溶液的pH值。

缓冲能力的大小与缓冲剂浓度、缓冲组分的比值有关，缓冲剂的浓度越大，缓冲能力越大,缓冲组分比值为1:1时缓冲能力最大。

缓冲溶液的pH值测定方法必须与计算方法相匹配。例如用pH试纸测定,则用(1)式计算即可;若用酸度计测定时,必须考虑活度,用(2)式计算。

3. 弱酸、弱碱及两性物质在水溶液中都发生质子传递反应。根据质子传递平衡理论,向该溶液中加入H^+或OH^-就可以阻止质子传递反应。另外由于弱酸(碱)的质子传递反应是吸热反应,所以加热可促进质子传递反应。

4. 难溶强电解质在一定温度下与它的饱和溶液中的相应离子处于平衡状态。例如:

$$AgCl(s) \rightleftharpoons Ag^+ + Cl^-$$

此时它的平衡常数称为溶度积K_{sp}。只要溶液中两种离子浓度乘积大于其溶度积,便有沉淀产生。反之如果降低饱和溶液中某种离子的浓度,使两种离子浓度的乘积小于其溶度积,则沉淀便会溶解。

【实验仪器与药品】

pHS-25型酸度计,烧杯(25ml,2只),量筒(10ml,3只),表面皿,甘汞电极,玻璃电极,复合电极。

HAc(0.1mol/L,1mol/L,2mol/L);$NH_3 \cdot H_2O$(0.1mol/L,1mol/L);甲基橙(0.1%);NH_4Cl(s);HCl(0.1mol/L,2mol/L,6mol/L);NaOH(2mol/L);NaAc(s);NaAc(0.1mol/L,1mol/L);pH试纸;甲基红(0.1%乙醇液);NaH_2PO_4(0.05mol/L);Na_2HPO_4(0.05mol/L);$SbCl_3$(0.2mol/L);$SnCl_2 \cdot 2H_2O$(s);酚酞(1%乙醇液);$MgCl_2$(0.1mol/L);$(NH_4)_2C_2O_4$(饱和);$CaCl_2$(0.1mol/L);NaCl(0.1mol/L);$AgNO_3$(0.1mol/L);磷酸盐标准缓冲液(pH 6.8)。

【实验内容】

1. 同离子效应

(1)在试管中加入约2ml 0.1mol/L HAc溶液,加1滴甲基橙指示液,观察溶液的颜色。然后加入少量固体NaAc,观察颜色有何变化,并解释之。

注:甲基橙:0.1%水溶液;变色范围pH 3.1~4.4(红→黄)。

(2)在试管中加入约2ml的0.1mol/L $NH_3 \cdot H_2O$溶液,加一滴酚酞指示液,观察溶液颜色。再加入少量固体NH_4Cl,观察颜色变化,并解释之。

注:酚酞指示液:1%乙醇溶液;变色范围pH 8.0~10.0(无色→红)。

2. 缓冲溶液的配制和性质

(1)向试管中加入10ml纯化水,再加3滴0.1mol/L HCl,摇匀后用pH试纸测其pH值。然后将溶液分成3份,一份加2滴2mol/L HCl,另一份加2滴2mol/L NaOH溶液,第三份是取原试管中HCl溶液1ml,加水至10ml,用pH试纸分别测定这三份溶液的pH值。

(2)在一试管中加入5ml的1mol/L HAc和5ml的1mol/L NaAc溶液,摇匀后,用pH试纸测其pH值,与理论值比较。将溶液分成3份,一份中加2滴2mol/L HCl溶液,另一份中加2滴2mol/L NaOH溶液,第三份取原试管中溶液1ml加水到10ml。分别测其pH值。与上实验溶液的pH值比较,由此得出什么结论?

(3) 欲配制 pH = 4.1 的缓冲液溶液 11ml，实验室现有 0.1mol/L HAc 和 0.1mol/L NaAc 溶液，应该怎样配制？先经过计算，再按计算的数量配好溶液，并用 pH 试纸测量其是否符合要求。

3. 缓冲能力　取 2 支试管，在 1 支试管中加入 0.1mol/L HAc 溶液和 0.1mol/L NaAc 溶液各 3ml，另 1 支试管中加入 1mol/L HAc 溶液和 1mol/L NaAc 溶液各 3ml，这时两试管溶液 pH 值是否相同？在两试管中分别滴入 2 滴甲基红指示剂，溶液呈红色（甲基红指示液在 pH < 4.2 时呈红色，pH > 6.3 时呈黄色），然后在两试管中分别逐滴加入 2mol/L NaOH 溶液（每加 1 滴摇匀），直至溶液的颜色变成黄色。记录各试管所加 NaOH 溶液的滴数。解释所得的结果。

4. 用 pHS – 25 型酸度计测量缓冲溶液的 pH 值　pHS – 25 型酸度计使用见附录。

(1) $0.05mol/L\ NaH_2PO_4\ 8ml$—$0.05mol/L\ Na_2HPO_4\ 2ml$ 缓冲液的配制；取 2 个干净并专用的 10ml 量筒，分别准确的量取 $0.05mol/L\ NaH_2PO_4\ 8ml$ 和 $0.05mol/L\ Na_2HPO_4\ 2ml$，倒入一预先洗净并干燥的 25ml 烧杯中，混合均匀，用 pHS – 25 酸度计测其 pH 值，与理论值对照。

(2) 用上述方法再配制一份 $0.05mol/L\ NaH_2PO_4\ 5ml$—$0.05mol/L\ Na_2HPO_4\ 5ml$ 缓冲溶液；用 pHS – 25 酸度计测其 pH 值，与理论值对照。

5. 用 pH 值试纸测试下列溶液（均为 0.1mol/L）的 pH 值　$NaCl$；NH_4Cl；NH_4Ac；Na_2CO_3；Na_3PO_4；Na_2HPO_4；NaH_2PO_4。

6. 盐类水解和影响水解平衡的因素

(1) 取 $0.2mol/L\ SbCl_3$ 溶液 5 滴，加 2ml 水稀释之，观察沉淀的产生 [$Sb(OH)_2Cl\downarrow \xrightarrow{-H_2O} SbOCl\downarrow$]，在沉淀上逐滴加入 6mol/L 盐酸，随加随振荡，观察沉淀是否溶解。如溶解，再加水稀释，沉淀是否又产生？解释以上各现象，写出 $SbCl_3$ 的质子传递反应式。

(2) 取少量 $SnCl_2 \cdot 2H_2O$ 晶体于试管中，加纯化水 1～2ml，观察有何现象发生？写出反应式，并解释之。

(3) 取试管 1 支，加入 1mol/L NaAc 溶液 2 滴，加 H_2O 至约 1ml，再加酚酞指示液 2～3 滴，加热至沸，再使之冷却，观察以上过程溶液颜色的变化，并解释之。

7. 沉淀的生成和溶解

(1) 在 2 支试管中分别加入 1ml 的 $0.1mol/L\ MgCl_2$ 溶液，并逐滴加入 $1mol/L\ NH_3 \cdot H_2O$ 至有白色 $Mg(OH)_2$ 沉淀生成，然后在第一支试管中加入 2mol/L HCl 溶液，沉淀是否溶解？在第二支试管中加入饱和 NH_4Cl 溶液，沉淀是否溶解？加入 HCl 和 NH_4Cl 对下列平衡各有何影响？

$$Mg(OH)_2 \rightleftharpoons Mg^{2+} + 2OH^-$$

(2) 在 2 支试管中分别加入 5 滴饱和 $(NH_4)_2C_2O_4$ 溶液和 5 滴 $0.1mol/L\ CaCl_2$ 溶液，观察白色 CaC_2O_4 沉淀的生成。然后在第一支试管内加入 2mol/L HCl 溶液约 1ml，搅拌，观察沉淀是否溶解？在另一支试管中加入 2mol/L HAc 溶液约 1ml，观察沉淀是否溶解？试加以解释。

(3) 取 0.5ml 的 $0.1mol/L\ NaCl$ 溶液，加几滴 $0.1mol/L\ AgNO_3$ 溶液，观察 AgCl 沉

淀的产生，再往其中加入 2mol/L $NH_3 \cdot H_2O$，观察由于配离子的生成而导致沉淀溶解。

Experiment 10 Acid – Base Equilibrium and Precipitation-Dissolution Equilibrium

Preview

1. Acid-base equilibrium and factors affecting the shift of the equilibrium.
2. The properties of buffer solution.
3. The formation and dissolution of precipitates.
4. The usage of pHS-25 acidometer.
5. The usage of pH test strips; the transfer of the liquid.

Questions

1. What are the similarities and differences between common ion effect and buffer solution?
2. How to inhibit or facilitate hydrolysis? Illustrate it.
3. Is it necessary under alkaline conditions to generate hydroxide precipitate? Is the pH of the solution same when the metal ion of different concentrations start to generate hydroxide precipitate?
4. Calculate the equilibrium constant of the following reaction:

(1) $Mg(OH)_2 + 2NH_4^+ \rightleftharpoons Mg^{2+} + 2NH_3 \cdot H_2O$

(2) $Fe(OH)_3 + 3NH_4^+ \rightleftharpoons Fe^{3+} + 3NH_3 \cdot H_2O$

(3) $Ag_2CrO_4 + 2Cl^- \rightleftharpoons 2AgCl + CrO_4^{2-}$

(4) $2AgCl + CrO_4^{2-} \rightleftharpoons Ag_2CrO_4 + 2Cl^-$

Principles

1. The ionization equilibrium of weak electrolytes and its shifting.

HA is a weak monobasic acid, and its ionization equilibrium in aqueous solution can be represented by:

$$HA \rightleftharpoons H^+ + A^-$$

Acidity constant is given by: $K_a = \dfrac{[H^+][A^-]}{[HA]}$

The conjugate base A^- is a weak monobasic base, its ionization equilibrium in aqueous solution can be represented by:

$$A^- + H_2O \rightleftharpoons HA + OH^-$$

Basicity constant is given by: $K_b = \dfrac{[HA][OH^-]}{[A^-]}$

A strong electrolyte containing the same ion was added into weak electrolyte HA aqueous solution, that is to say increase A^- or H^+ ion concentration, the equilibrium will shift to form

HA molecules, and then the ionization of weak electrolyte HA is suppressed. This effect is called as the common ion effect.

2. Buffer solution is generally consisted of the conjugate acid-base pair. The pH value can be calculated by the following formula:

$$pH = pK_a + \lg(c_b/c_a) \tag{1}$$

Considering the activity, the pH value can be calculated as follows:

$$pH = pK_a + \lg\frac{f_2 c_b}{f_1 c_a} \tag{2}$$

Here, f_1, f_2 are activity coefficient of acids and base respectively. Excluding the activity, error between calculated and experimental data will be introduced by using concentration instead of activity.

The pH of the buffer solution is mainly determined by pK_a, but also is related to the ratio of concentration of base and acid. If original concentrations of base and acid are same, the radio of volume of base and acid is equal to that of concentration ratio. So equation (1) can be rewritten as:

$$pH = pK_a + \lg(V_b/V_a)$$

The buffer solution with the desired pH value will be obtained by changing the volume ratio of base and acid solutions.

During the process of diluting a buffer solution, the concentrations of base and acid are reduced in equal proportions, the concentration ratio of base and acid will not change. So a suitable dilution does not affect the pH of buffer solution.

Buffering capacity depends on buffer concentration and the ratio of buffer components. The buffer capacity increases with the buffer concentration; when buffer component ratio is 1:1, buffer capacity reach the max value.

Calculation of the pH value of buffer solution depends on the determination of pH value. For example, measuring by pH test strips, then the pH value can be calculated by (1); when measuring with a pH acidometer, then the pH value can be calculated by (2) considering activity.

3. Weak acid, weak base and amphoteric substances all have proton transfer reactions in aqueous. According to proton transfer equilibrium theory, increasing H^+ or OH^- concentration can suppress the proton transfer reaction. In addition, the proton transfer reaction of weak acid (base) is endothermic reaction, so heating can promote the proton transfer reactions.

4. Insoluble strong electrolyte and its corresponding ions are in balance in a saturated solution at certain temperature. For example:

$$AgCl(s) \rightleftharpoons Ag^+ + Cl^-$$

Its equilibrium constant is called solubility product constant K_{sp}. As long as both ions concentration product in the solution is greater than the solubility product, it will be precipitated. Conversely, if reducing one ion concentration in a saturated solution, the ions concentration product is less than the solubility product, the precipitate will be dissolved.

Instruments and Chemicals

pHS-25 acidometer, beakers (25ml × 2), measuring cylinders (10ml × 3), calomel electrode, glass electrode, combination electrode.

HAc (0.1mol/L, 1mol/L, 2mol/L), $NH_3 \cdot H_2O$ (0.1mol/L, 1mol/L), Methyl orange indicator solution (0.1%), NH_4Cl (s), HCl (0.1mol/L, 2mol/L, 6mol/L), NaOH (2mol/L), NaAc (s), NaAc (0.1mol/L, 1mol/L), pH test strips, Methyl red indicator solution (0.1% ethyl alcohol), NaH_2PO_4 (0.05mol/L), Na_2HPO_4 (0.05mol/L), $SbCl_3$ (0.2mol/L), $SnCl_2 \cdot 2H_2O$ (s), phenolphthalein indicator (1% ethyl alcohol), $MgCl_2$ (0.1mol/L), $(NH_4)_2C_2O_4$ (saturated), $CaCl_2$ (0.1mol/L), NaCl (0.1mol/L), $AgNO_3$ (0.1mol/L), phosphate standard buffer solution (pH 6.8).

Procedures

1. Common ion effect

(1) Add about 2ml 0.1mol/L HAc solution to a tube, add one drop of Methyl orange indicator solution, and observe the color of solution. Then add few NaAc solid, note the color change of solution and explain it.

Tips: Methyl orange solution: 0.1% aqueous solution; the range of the color change is pH 3.1—4.4 (red → yellow).

(2) Add about 2ml 0.1mol/L $NH_3 \cdot H_2O$ solution, then add one drop of phenolphthalein indicator, observe the color of solution. Then add few NH_4Cl solid, Note the color change of solution and explain it.

Tips: phenolphthalein indicator: 1% ethyl alcohol, the range of the color change is pH 8.0—10.0 (colorless → red).

2. The preparation and properties of buffer solution

(1) Add 10ml water to a tube, then add 3 drops of 0.1mol/L HCl solution to the second tube, shake well and test pH with pH test strips. Divide the solution into three portions, add 2 drops of 2mol/L HCl solution to the first solution; add 2 drops of 2mol/L NaOH solution to the second tube, take 1ml solution from the original tubes and add this solution into the third tube, then add water until the solution volume is 10ml. Determine the pH value with pH test strips.

(2) Add 5ml 1mol/L HAc and 5ml 1mol/L NaAc solution to a tube, and shake well, then test the pH value of the solution with pH test strips comparing with theoretical value. Divide the solution into three portions, add 2 drops of 2mol/L HCl solution to the first tube, add 2 drops of 2mol/L NaOH solution to the second tube, take 1ml solution from the original tubes to the third tube and add water until the solution volume is 10ml. Determine the pH value with pH test strips.

(3) How to prepare a buffer solution (pH = 4.1) 11ml with 0.1mol/L HAc and 0.1mol/L NaAc solution? Calculate it and prepare the solution through the calculation, then

verify the solution with pH test strips.

3. Buffer capacity

Add 3ml 0.1mol/L HAc solution and 3ml 0.1mol/L NaAc solution to one tube, and add 3ml 1mol/L HAc solution and 3ml 1mol/L NaAc solution to the other. Is the pH of two solutions equal? Then add two drops of methyl red indicator solution to each tube, the color of solutions are red (Methyl red indicator solution: pH < 4.2 red, pH > 6.3 yellow), then add 2mol/L NaOH solution dropwise with shaking until the color of solution turns yellow. Record how many drops of NaOH are used. Explain the results.

4. Determine the pH value of buffer solution with pHS-25 acidometer

The instructions of pHS-25 acidometer are shown in appendix.

(1) The preparation of buffer solution (0.05mol/L NaH_2PO_4 8ml—0.05mol/L Na_2HPO_4 2ml): take two clean and special 10ml measuring cylinder, measure 8ml 0.05mol/L NaH_2PO_4 with one cylinder precisely, measure 2ml 0.05mol/L Na_2HPO_4 with another cylinder precisely, decant these solutions into a dry and clean 25ml beaker, mix well and determine the pH value with pHS-25 acidometer, compare with the calculated value.

(2) Prepare buffer solution (0.05mol/L NaH_2PO_4 5ml—0.05mol/L Na_2HPO_4 5ml): with the method in test (1), determine the pH value with pHS-25 acidometer, compare with the calculated value.

5. Determine the pH value of the following solution (0.1mol/L)

$NaCl$; NH_4Cl; NH_4Ac; Na_2CO_3; Na_3PO_4; Na_2HPO_4; NaH_2PO_4.

6. Hydrolysis of salts and influencing factors of hydrolysis equilibrium

(1) Add 5 drops of 0.2mol/L $SbCl_3$ solution to a tube, add 2ml water, observe the emergence of precipitation ($Sb(OH)_2Cl \downarrow \xrightarrow{-H_2O} SbOCl \downarrow$), and add several drops of 6mol/L HCl dropwise with shaking. Observe whether the precipitation dissolves. If the precipitation dissolves, dilute the solution with water, does the precipitation emerge? Explain the phenomenon and write down the proton transfer equation of $SbCl_3$.

(2) Add litter amount of crystal $SnCl_2 \cdot 2H_2O$ to a tube, then add 1—2ml H_2O, what happen? Write down the reaction equation and explain it.

(3) Add 2 drops of 1mol/L NaAc solution and 1ml water to a tube, then add 2—3 drops of phenolphthalein indicator, heat to boil, then cool down. Observe the color change of solution and explain it.

7. Formation and dissolution the precipitation

(1) In two tubes, both add 1ml 0.1mol/L $MgCl_2$ solution and add 1mol/L $NH_3 \cdot H_2O$ dropwise until the emergence of white precipitation $Mg(OH)_2$, then add 2mol/L HCl solution to one tube, does the precipitation dissolve? Add saturated NH_4Cl solution to another tube, does the precipitation dissolves? What is the influence to the following balance when add HCl or NH_4Cl?

$$Mg(OH)_2 \rightleftharpoons Mg^{2+} + 2OH^-$$

(2) In two tubes, both mix 5 drops of saturated $(NH_4)_2C_2O_4$ and 5 drops of 0.1mol/L $CaCl_2$ solution; observe the emergence of white CaC_2O_4 precipitation. Then add 1ml 2mol/L HCl solution to one tube, does the precipitation dissolves? add 1ml 2mol/L HAc solution to another tube, does the precipitation dissolves? Try to explain it.

(3) Take 0.5ml 0.1mol/L NaCl solution, add several drops of 0.1mol/L $AgNO_3$ solution, and observe the emergence of AgCl precipitation. Then add 2mol/L $NH_3 \cdot H_2O$, observe the dissociation of precipitation for generation of complex ion.

Vocabulary

buffer solution	缓冲溶液	pH test strips	pH 试纸
common ion effect	同离子效应	hydrolysis	水解作用
concentration	浓度	buffering capacity	缓冲能力
Methyl orange indicator solution	甲基橙指示剂		
Methyl red indicator solution	甲基红指示剂		
phenolphthalein indicator	酚酞指示剂		

（编写：段丽颖，英文核审：夏丹丹）

实验十一　氧化还原反应

【预习内容】

1. 用 pHS-25 型酸度计测量电池的电动势。
2. 原电池、电极反应、电池反应、电池符号。
3. 浓度对电极电势的影响。
4. 电极电势的应用。

【思考题】

1. 影响电极电势的因素有哪些？
2. 浓度、酸度对氧化还原产物及反应方向有何影响？
3. 盐桥有什么作用？
4. （标准）电极电势有何应用？

【实验原理】

氧化还原反应是物质得失电子的过程。

电极电势是判断氧化剂和还原剂相对强弱的标准，并可用以确定氧化还原反应进行的方向。电极电势表是各物质在水溶液中进行氧化还原反应规律性的总结，溶液的浓度、温度均影响电极电势的数值。一般来说，在表中电极电势数值小的还原型是较强的还原剂，电极电势数值大的氧化型是较强的氧化剂。

$$氧化型 + ne \rightleftharpoons 还原型$$

氧化还原反应的方向是电极电势高的氧化型物质与电极电势低的还原型物质反应。

在一定条件下，将氧化还原反应所放出的能量转变成电能的装置称为原电池。
例：
$$Zn + CuSO_4\ (a_2) \rightleftharpoons ZnSO_4\ (a_1) + Cu$$
其装置如图 11-1。其电池组成为：
$$(-)Zn|Zn^{2+}(a_1) \| Cu^{2+}(a_2)|Cu(+)$$

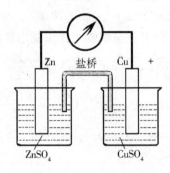

图 11-1　丹尼尔电池（铜-锌电池）装置图

电池电动势主要决定于锌的电极电势 $\varphi(Zn^{2+}/Zn)$ 和铜的电极电势 $\varphi(Cu^{2+}/Cu)$，电池的电动势为：$E = \varphi(Cu^{2+}/Cu) - \varphi(Zn^{2+}/Zn)$

若已知电极液活度，电动势可用 Nernst 公式计算：
$E = \varphi(+) - \varphi(-)$，当 $T = 298K$ 时：
$$E = [\varphi^{\ominus}(Cu^{2+}/Cu) + \frac{0.0592}{n}\lg\frac{a(Cu^{2+})}{a(Cu)}] - [\varphi^{\ominus}(Zn^{2+}/Zn) + \frac{0.0592}{n}\lg\frac{a(Zn^{2+})}{a(Zn)}]$$
$$= [\varphi^{\ominus}(Cu^{2+}/Cu) - \varphi^{\ominus}(Zn^{2+}/Zn)] + \frac{0.0592}{n}\lg\frac{a(Cu^{2+}) \cdot a(Zn)}{a(Zn^{2+}) \cdot a(Cu)}$$

因为 $a_{Cu} = 1$，$a_{Zn} = 1$ 所以上式可化简为：
$$E = [\varphi^{\ominus}(Cu^{2+}/Cu) - \varphi^{\ominus}(Zn^{2+}/Zn)] + \frac{0.0592}{n}\lg\frac{a(Cu^{2+})}{a(Zn^{2+})}$$

电极电势的测定：所谓电极电势是指某电对以标准氢电极为基准而得出的该电对的相对平衡电势，在此原则上只要将待测电极与标准氢电极（或其他参比电极）组成电池，则其电动势（E）为：
$$E = \varphi(+) - \varphi(-) \tag{1}$$

由于其中一个电对（为标准氢电极或参比电极）的电极电势为已知，所以测此原电池的电动势通过（1）式即可求得未知电对的电极电势。

在实际测定电极电势的工作中，由于标准氢电极控制的条件很严，使用时不方便，目前多采用一些工艺简单、制作方便、电极电势稳定的电极作参比电极来代替标准氢电极。常用的参比电极有甘汞电极。它是由 Hg、Hg_2Cl_2（s）及 KCl 溶液组成。其电极反应为：
$$Hg_2Cl_2\ (s) + 2e = 2Hg\ (l) + 2Cl^-$$
$$\varphi = \varphi^{\ominus} + \frac{RT}{2F}\ln\frac{1}{[Cl^-]^2}$$

因此甘汞电极的电势与 KCl 溶液的浓度有关。温度对电极电势也有一定影响。最常用的是饱和甘汞电极。因此，可用饱和甘汞电极来代替标准氢电极。298K 时饱和甘汞电极电势 $\varphi = 0.2415\text{V}$，标准甘汞电极电势 $\varphi^{\ominus} = 0.2828\text{V}$。

例如：测定铜的电极电势可组成下列电池：

$$(-)\ \text{Hg}\ |\ (1)\ \text{Hg}_2\text{Cl}_2\ (s),\ \text{KCl}（饱和）\ \|\ \text{Cu}^{2+}\ (a)\ |\ \text{Cu}\ (+)$$

测其电动势为 $E = \varphi(\text{Cu}^{2+}/\text{Cu}) - \varphi(甘汞)$，所以 $\varphi(\text{Cu}^{2+}/\text{Cu}) = E + \varphi$ (甘汞)，若已知铜电极的溶液活度，可根据下式由铜电极电势求得铜的标准电极电势：

$$\varphi(\text{Cu}^{2+}/\text{Cu}) = \varphi^{\ominus}(\text{Cu}^{2+}/\text{Cu}) + \frac{0.0592}{2}\lg a(\text{Cu}^{2+})$$

$$\varphi^{\ominus}(\text{Cu}^{2+}/\text{Cu}) = \varphi(\text{Cu}^{2+}/\text{Cu}) - \frac{0.0592}{2}\lg a(\text{Cu}^{2+})$$

由文献已知：298K 时 0.5mol/L $\text{CuSO}_4\ f_{\pm} = 0.067$；0.5mol/L $\text{CuSO}_4\ f_{\pm} = 0.069$。

【实验仪器与药品】

pHS-25 型酸度计；试管（离心，10ml）；烧杯（5ml，2 只）；饱和甘汞电极；盐桥。

KMnO_4（0.01mol/L，饱和）；H_2SO_4（2mol/L）；H_2O_2（3%）；FeSO_4（0.1mol/L）；$\text{K}_2\text{Cr}_2\text{O}_7$（0.1mol/L）；$\text{H}_2\text{S}$（饱和）；KI（0.1mol/L，1mol/L）；FeCl_3（0.1mol/L）；CCl_4；KBr（0.1mol/L，1mol/L）；碘水（0.1mol/L）；HCl（浓，0.01mol/L）；$(\text{NH}_4)_2\text{Fe}(\text{SO}_4)_2\cdot 6\text{H}_2\text{O}$（s）；$\text{AgNO}_3$（0.1mol/L）；$\text{NH}_4\text{SCN}$（10%）；$\text{K}_3[\text{Fe}(\text{CN})_6]$（0.1mol/L）；$\text{ZnSO}_4$（0.1mol/L）；$\text{Na}_2\text{SO}_3$（1.0mol/L）；NaOH（6mol/L）；$\text{MnSO}_4$（0.2mol/L）；$(\text{NH}_4)_2\text{S}_2\text{O}_8$（s）；$\text{ZnSO}_4$（0.5mol/L）；$\text{CuSO}_4$（0.5mol/L）。

材料：电极（锌棒，铜棒），淀粉 KI 试纸，导线，砂纸，滤纸。

【实验内容】

1. 氧化剂和还原剂

(1) 取 2 支试管，各加 0.01mol/L KMnO_4 溶液 5 滴及 6mol/L H_2SO_4 3 滴，然后在第一支试管中加入 3% H_2O_2 溶液 1 滴，第二支试管中加入 0.1mol/L FeSO_4 溶液 2~3 滴，观察现象，写出反应式，并指出反应中的氧化剂和还原剂。

(2) 取 1 支试管，加入 0.1mol/L $\text{K}_2\text{Cr}_2\text{O}_7$ 溶液 3 滴，6mol/L H_2SO_4 5 滴，再加 H_2S 水溶液数滴，摇匀，观察现象。写出反应式，并指出反应中的氧化剂和还原剂。

2. 氧化还原反应和电极电势

(1) 在试管中加入 0.5ml 0.1mol/L KI 溶液和 2 滴 0.1mol/L FeCl_3 溶液，摇匀后加入 0.5ml CCl_4，充分振荡，观察 CCl_4 层有无变化。

(2) 用 0.1mol/L KBr 溶液代替 KI 溶液进行同样实验，观察现象。

(3) 往两支试管中分别加入 3 滴碘水、溴水，然后加入约 0.5ml 0.1mol/L FeSO_4 溶液，摇匀后，滴入 0.5ml CCl_4，充分振荡，观察 CCl_4 层有无变化。

根据上实验结果，定性地比较 Br_2/Br^-、I_2/I^- 和 $\text{Fe}^{3+}/\text{Fe}^{2+}$ 3 个电对的电极电势。

3. 氧化型、还原型浓度和介质条件对氧化还原反应的影响

(1) 酸度对氧化还原反应的影响　取 2 支试管（尽量控干），分别加入饱和 KMnO_4

溶液各 1 滴，然后分别滴入浓 HCl（12mol/L）和 0.01mol/L HCl 各约 1ml，观察现象，并用润湿的淀粉碘化钾试纸检验是否有 Cl_2 气产生，为什么？试根据 Nernst 方程解释之。

$[H^+]$ mol/L	$\varphi(MnO_4^-/Mn^{2+})$ (V)	$[Cl^-]$ mol/L	$\varphi(Cl_2/Cl^-)$ (V)	现象（淀粉 KI 试纸的变化）	结论

(2) 沉淀对氧化还原反应的影响

①取少量硫酸亚铁铵 $[(NH_4)_2SO_4 \cdot FeSO_4 \cdot 6H_2O]$ 固体加入离心试管中，加入少量水使之溶解，再加入 1~2 滴碘试液，混匀后观察碘试液颜色是否褪去？然后向离心试管中滴加 0.1mol/L $AgNO_3$ 溶液，边加边振荡，注意碘的棕黄色是否褪去。离心沉降，吸取上清液转移至另一支试管中，往清液中加几滴 10% 的 NH_4SCN 溶液，观察颜色变化，解释现象并写出各步相应的反应方程式。

②往试管中加入 5 滴 0.1mol/L KI 溶液，2 滴 0.1mol/L $K_3[Fe(CN)_6]$ 溶液，混匀后，再加入 0.5ml CCl_4，充分振荡，观察 CCl_4 层的颜色有无变化？然后加入 2 滴 0.1mol/L 的 $ZnSO_4$ 溶液，充分振荡，观察现象并加以解释。根据 $\varphi^{\ominus}$ 值判断 I^- 是否能还原 $[Fe(CN)_6]^{3-}$，加入 Zn^{2+} 有何影响？

提示：

$$2I^- + 2[Fe(CN)_6]^{3-} = I_2 + 2[Fe(CN)_6]^{4-}$$
$$+ Zn^{2+}$$
$$\rightarrow Zn_2[Fe(CN)_6]\downarrow（白色）$$

(3) 介质条件对氧化还原反应产物的影响　取 3 支试管，各加入 0.01mol/L $KMnO_4$ 溶液 3 滴，再分别加入 6mol/L H_2SO_4、纯化水和 6mol/L NaOH 溶液各 0.5ml，摇匀后，再各加入 1mol/L Na_2SO_3 溶液 1ml，观察现象，写出反应。

4. 催化剂对氧化还原反应的影响　取 5ml 的 0.2mol/L $MnSO_4$ 溶液和 1ml 的 6mol/L H_2SO_4 溶液在试管内充分振摇，并加入一小匙过二硫酸铵固体，充分振摇溶解后分成 2 份，往一份溶液中加 1~2 滴 0.1mol/L $AgNO_3$ 溶液，静止片刻，观察溶液颜色有何变化？与未加 $AgNO_3$ 溶液的份溶液比较，反应情况有何不同？

5. 选择氧化剂　为了使 I^- 氧化为 I_2，而又不使 Br^-、Cl^- 被氧化，在常用的氧化剂 $FeCl_3$ 和 $KMnO_4$ 中选择哪一种符合要求？自行设计方案。

已知实验室中有 1mol/L NaCl、KBr、KI 溶液。

提示：应考虑液性。如何观察反应是否产生？

6. 铜锌电池电动势测定　在 2 支 50ml 的小烧杯中，分别加入 25ml 的 0.5mol/L $ZnSO_4$ 和 25ml 的 0.5mol/L $CuSO_4$ 溶液，在 $ZnSO_4$ 溶液中插入锌棒，$CuSO_4$ 溶液中插入铜棒，组成 2 个电极，中间以盐桥相通。用 pHS-25 型酸度计测其电动势（见附录）。

7. 电极电势的测定

$$(-) Zn | ZnSO_4 (0.5mol/L) \| KCl (饱和), Hg_2Cl_2 (s) | Hg (+)$$

取饱和甘汞电极和锌电极（0.5mol/L）组成原电池，用 pHS-25 型酸度计测其动势。由所测电动势计算锌电极的电极电势，并由此计算锌电极的标准电极电势（实验值），与理论值比较。

附注：盐桥的制法

称取 1g 琼脂，放在 100ml KCl 饱和溶液中浸泡一会儿，在不断搅拌下，加热煮成糊状，趁热倒入 U 形玻璃管中（管内不能留有气泡，否则会增加电阻），冷却即成。

更为简便的方法可用 KCl 饱和溶液装满 U 形玻璃管，两管口以小棉花球塞住（管内不留气泡），作为盐桥使用。

实验中还可用素烧瓷筒用作盐桥。电极的处理：电极的锌棒、铜棒要用砂纸擦干净，以免增大电阻。

Experiment 11 Redox Reactions

Preview

1. Measuring the cell potential with pHS-25 acidometer.
2. Galvanic battery, electrode reaction, cell reaction, cell diagram.
3. The effect of concentration on the electrode potential.
4. Applications of electrode potentials.

Questions

1. What factors will influence on electrode potential?
2. How do concentrations and acidity affect redox product and direction of the reaction?
3. What is the role of salt bridge?
4. What are applications of standard electrode potential?

Principles

Redox reaction is a process which occurring by the loss of electrons from one substance and gain by another substance.

Electrode potential is the criterion for judging oxidant and reductant relative strength, and can also be used to predict the direction of the redox reaction. Electrode potential tables is a summary that each substance in aqueous solution takes place redox reaction, the concentration and temperature both affect the value of the electrode potential. Generally speaking, in the table the value of the electrode potential is smaller, reduction form is a stronger reducing agent; the value of the electrode potential is larger, oxidation form is a stronger oxidizing agent.

$$\text{oxidation form} + ne \rightleftharpoons \text{reduction form}$$

The redox reaction takes place between oxidation form material with high electrode potential and reduction form material with low electrode potential.

In certain conditions, the device designed to produce electron current basing on redox reaction is called galvanic cell.

For example,
$$Zn + CuSO_4(a_2) \rightleftharpoons ZnSO_4(a_1) + Cu$$

Daniell cell is illustrated in Fig. 11-1. The cell diagram is written as:
$$(-)Zn|Zn^{2+}(a_1) \| Cu^{2+}(a_2)|Cu(+)$$

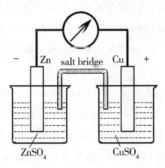

Fig. 11-1 the schematric of Daniell cell

The cell potential is the difference between the electrode potentials of the two half-cells:
$$E = \varphi(Cu^{2+}/Cu) - \varphi(Zn^{2+}/Zn)$$

If the activity of electrode solution is given, the cell potential can be calculated by Nernst equation:
$$E = \varphi(+) - \varphi(-)$$

$$E = (\varphi^{\ominus}(Cu^{2+}/Cu) + \frac{0.0592}{n}\lg\frac{a(Cu^{2+})}{a(Cu)}) - (\varphi^{\ominus}(Zn^{2+}/Zn) + \frac{0.0592}{n}\lg\frac{a(Zn^{2+})}{a(Zn)})$$

$$= (\varphi^{\ominus}(Cu^{2+}/Cu) - \varphi^{\ominus}(Zn^{2+}/Zn)) + \frac{0.0592}{n}\lg\frac{a(Cu^{2+}) \cdot a(Zn)}{a(Zn^{2+}) \cdot a(Cu)} \text{ where } T = 298K$$

The activities of Zn and Cu metal are taken to be unity ($a_{Cu} = 1, a_{Zn} = 1$), so the above equation can be simplified as:
$$E = (\varphi^{\ominus}(Cu^{2+}/Cu) - \varphi^{\ominus}(Zn^{2+}/Zn)) + \frac{0.0592}{n}\lg\frac{a(Cu^{2+})}{a(Zn^{2+})}$$

The determination of electrode potential:

The standard reduction potential is that the potential of an unknown electrode is universally expressed relative to that of standard hydrogen electrode. On this principle, it is possible to combine the test electrode with the standard hydrogen electrode (or other reference electrode), measure the cell potential (E), and, thus, to obtain the electrode potential of the test electrode by the following equation:
$$E = \varphi(+) - \varphi(-) \qquad (1)$$

In routine experimental work, the standard hydrogen electrode is inconvenient to handle. A reference electrode, which is easily manufactured and handled, and well controlled, is required in the laboratory. A common reference electrode is calomel electrode, which consists of Hg, $Hg_2Cl_2(s)$ and KCl solution. The half-cell reaction is:

$$Hg_2Cl_2(s) + 2e = 2Hg(l) + 2Cl^-$$

$$\varphi = \varphi^{\ominus} + \frac{RT}{2F}\ln\frac{1}{[Cl^-]^2}$$

Therefore, calomel electrode potential is related to the concentration of the KCl solution and temperature of system. Here, a most popular reference electrode, saturated calomel electrode is used to be instead of the standard hydrogen electrode. The potential of saturated calomel electrode is 0.2415V and the potential of standard calomel electrode is 0.2828V at 298K.

For example: the potential of Cu^{2+}/Cu half-cell can be determined by combining a galvanic cell as follows:

$$(-) Hg|(l)Hg_2Cl_2(s), KCl(saturated) \| Cu^{2+}(a)|Cu(+)$$

Cell potential, $E = \varphi(Cu^{2+}/Cu) - \varphi(\text{calomel electrode})$, is measured, therefore, the potential of Cu^{2+}/Cu half-cell can be obtained via $\varphi(Cu^{2+}/Cu) = E + \varphi(\text{calomel electrode})$. If solution activity of Cu^{2+}/Cu half-cell is known, according to the following formula, standard copper electrode potential can be obtained via the copper electrode potential:

$$\varphi(Cu^{2+}/Cu) = \varphi^{\ominus}(Cu^{2+}/Cu) + \frac{0.0592}{2}\lg a(Cu^{2+})$$

$$\varphi^{\ominus}(Cu^{2+}/Cu) = \varphi(Cu^{2+}/Cu) - \frac{0.0592}{2}\lg a(Cu^{2+})$$

Data from the literature: For 0.5mol/L $CuSO_4$ $f_{\pm} = 0.067$; for 0.5mol/L $ZnSO_4$ $f_{\pm} = 0.069$ at 298K.

Instrument and Chemicals

pHS-25 acidometer, centrifuge tube (10ml), beakers (5ml × 2), saturated calomel electrode, salt bridge.

$KMnO_4$(0.01mol/L, saturated); H_2SO_4(2mol/L); H_2O_2(3%); $FeSO_4$(0.1mol/L); $K_2Cr_2O_7$(0.1mol/L); H_2S(saturated); $FeCl_3$(0.1mol/L); CCl_4; KBr (0.1mol/L, 1mol/L); KI (0.1mol/L, 1mol/L); iodine solution (0.1mol/L); HCl (saturated, 0.01mol/L); $(NH_4)_2Fe(SO_4)_2 \cdot 6H_2O$(s); $AgNO_3$(0.1mol/L); NH_4SCN(10%); $K_3[Fe(CN)_6]$ (0.1mol/L); $ZnSO_4$(0.1mol/L, 0.5mol/L); Na_2SO_3(1.0mol/L); NaOH (6mol/L); $MnSO_4$ (0.2mol/L); $(NH_4)_2S_2O_8$(s); $CuSO_4$(0.5mol/L).

Material

electrode (zinc rod, copper rod), KI-containing starch detection, wire, sand paper, filter paper.

Procedures

1. Oxidant and reductant

(1) In each of two tubes, mix 5 drops of 0.01mol/L $KMnO_4$ solution with 3 drops of 6mol/L H_2SO_4. To one tube, add 1 drop of 3% H_2O_2, and to the other tube, add 2—3 drops of 0.1mol/L $FeSO_4$ solution. Note your observation, write down equations for those processes and identify what the oxidant is and what the redundant is.

(2) Add 3 drops of 0.1mol/L $K_2Cr_2O_7$ solution and 5 drops of 6mol/L H_2SO_4 solution to

one tube, and then add a few drops of H_2S solution. Note your observation after shaking. Write down the equation and identify what the oxidizing agent is and what the reducing agent is.

2. Redox reaction and electrode potential

(1) Add 0.5ml of 0.1mol/L KI solution and two drops of 0.01mol/L $FeCl_3$ solution to a tube, shake well and add 0.5ml CCl_4, stir and note the change of CCl_4 layer after shaking.

(2) Repeat test (1) with 0.1mol/L KBr instead of KI solution. Note your observation.

(3) To one tube, add 3 drops of iodine aqueous solution. To the other, add 3 drops of bromine aqueous solution. To each of the two tubes, add 0.5ml of 0.1mol/L $FeSO_4$ solution, shake well and add 0.5ml of CCl_4, and note the change of CCl_4 layer after shaking.

According to the above experimental results, qualitatively compare the electrode potentials of redox couples Br_2/Br^-, I_2/I^- and Fe^{3+}/Fe^{2+}.

3. The effect of concentration and medium conditions on redox reactions

(1) The effect of acidity on redox reaction

To each of two tubes, add one drop of saturated $KMnO_4$ solution, then to one tube add 1ml of 12mol/L HCl, To another tube, add 1ml of 0.01mol/L HCl. Observe the phenomenon and use KI-containing starch detection to verify whether the gas of Cl_2 produces. Try to explain it according to Nernst equation.

$[H^+]$ mol/L	$\varphi(MnO_4^-/Mn^{2+})(V)$	$[Cl^-]$ mol/L	$\varphi(Cl_2/Cl^-)(V)$	Phenomenon (the change of KI - containing starch detection)	conclusion

(2) The effect of precipitation on redox reactions

①Add few ammonium ferrous sulphate (($NH_4)_2SO_4 \cdot FeSO_4 \cdot 6H_2O$) solid to a centrifugal tube, add few water to make solid dissolved, and then add 1—2 drops of iodine aqueous solution. Observe whether iodine solution fade after the solution is mixed well. Add several drops of 0.1mol/L $AgNO_3$ solution to the centrifugal tube, pay attention to the color change of iodine solution after shaking. Centrifuge, divert supernatant to another tube, and add several drops of 10% NH_4SCN solution, note the color change, explain the phenomenon and write down the corresponding equations.

②Add 5 drops of 0.1mol/L KI solution and 2 drops of 0.1mol/L $K_3[Fe(CN)_6]$ solution, mix well, then add 0.5ml of CCl_4, shake well, and observe the color change of CCl_4 layer. Then add 2 drops of 0.1mol/L $ZnSO_4$ solution, shake well, Note and explain your observation. According to the value of $\varphi^\ominus$, predict whether $[Fe(CN)_6]^{3-}$ can be reduced by I^-. What will happen after adding Zn^{2+} to the system?

Tips:

$$2I^- + 2[Fe(CN)_6]^{3-} = I_2 + 2[Fe(CN)_6]^{4-}$$
$$+ Zn^{2+}$$
$$\rightarrow Zn_2[Fe(CN)_6] \downarrow (white)$$

(3) The effect of medium on redox reactions.

In each of three tubes, add 3 drops of 0.01mol/L $KMnO_4$. Then to the first tube, add 0.5ml of 6mol/L H_2SO_4. To the second tube, add 0.5ml of water. To the third tube, add 0.5ml of 6mol/L NaOH solution. Add 1ml of 1mol/L Na_2SO_3 solution to every tube. Note your observation and write down the corresponding equations.

4. The effect of catalyst on redox reactions

Mix 5ml of 0.2mol/L $MnSO_4$ solution and 1ml of 6mol/L H_2SO_4 solution, then add a spoon of $(NH_4)_2S_2O_8$ solid. Shake constantly to dissolve the solid completely. Divide the solution into two portions. To one portion, add 1—2 drops of 0.1mol/L $AgNO_3$ solution and let it sit for a few minutes. Compare the color change of solution with that of solution without $AgNO_3$.

5. Selection of oxidizing agents

Between the common used oxidizing agents, $FeCl_3$ and $KMnO_4$, choose an oxidizing agent which can oxidize I^- to I_2 while not oxidize Br^- and Cl^-. Design the experiment.

1mol/L NaCl, 1mol/L KBr and 1mol/L KI solution are available in laboratory.

Tips: considering medium. How to prove your experimental results?

6. Determination of the cell potential difference of Daniell cell

Add 25ml of 0.5mol/L $ZnSO_4$ and 25ml of 0.5mol/L $CuSO_4$ to two 50ml bakers respectively. A Zn stripe immersed in the $ZnSO_4$ solution and a Cu stripe immersed in the $CuSO_4$ solution are combined in an electrical circuit. The two solutions in this galvanic cell are connected by a salt bridge. Determine the value of cell potential by pHS-25 acidometer. (Show in appendix)

7. Determine electrode potential

(−) Zn|$ZnSO_4$(0.5mol/L) ∥ KCl(saturated), Hg_2Cl_2(s)|Hg (+)

Combine saturated calomel electrode and Zn electrode to form a galvanic cell, use pHS-25 acidometer to test cell potential. According to the value of cell potential in this test, calculate the electrode potential of standard Zn electrode, and compare it with the theoretical value.

Remark: The preparation of salt bridge

Weigh and place 1g of agar in 100ml of KCl saturated solution, immersed for a few minutes. Then boil the mixture to a paste with constantly stirring. Pour into a U-shaped glass tube (Make sure no bubbles in the tube to avoid increasing the resistance). A salt bridge is done after cooling to room temperature.

A more convenient method to make a salt brigde is that KCl saturated solution is filled in U-shaped glass tube, two ports with small cotton balls stuffed (no bullbes appear in the tube).

An unglazed porcelain tube can also be used as salt bridge.

Electrode pre-treatment: zinc electrode, copper electrode must be polished by sandpaper in order to avoid increasing the resistance.

Vocabulary

galvanic cell	原电池	Daniell cell	铜锌电池
cell diagram	电池符号	cell potential	电池电动势
oxidant	氧化剂	reductant	还原剂
calomel electrode	甘汞电极	unglazed porcelain	素瓷

（编写：段丽颖，英文核审：夏丹丹）

实验十二 分光光度法测定磺基水杨酸合铁（Ⅲ）配合物的组成和稳定常数

【预习内容】
1. 分光光度计的用法。
2. 分光光度法测定配合物的组成和稳定常数的原理。

【思考题】
1. 为什么配合物的生成会受到溶液酸度影响？
2. 在实验中，每个溶液的 pH 值是否相同？如果不同，如何影响实验结果？
3. 用分光光度法测定配合物的组成和稳定常数的前提条件是什么？

【实验原理】

磺基水杨酸（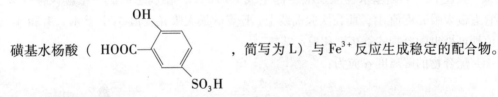，简写为 L）与 Fe^{3+} 反应生成稳定的配合物。配合物的颜色和组成随溶液的 pH 变化而变化。pH<4，形成紫色的配合物 $[FeL(H_2O)_4]^+$；pH 在 4~8，形成红色 $[FeL_2(H_2O)_2]^-$；pH 在 10 左右，形成黄色 $[FeL_3]^{3-}$。在本实验中，测定的是 pH 在 2~3 时紫色配合物的组成和稳定常数。

根据朗伯-比尔定律，当一束单色光通过有色溶液时，溶液的吸光度与吸光物质的浓度呈正比关系。即：

$$A = \varepsilon l c$$

式中，A 为溶液的吸光度；c 为吸光物质浓度（mol/L）；l 为液层厚度（cm）；ε 为摩尔吸光系数，当给定单色光、溶剂和温度等条件时，它是物质的特征常数。

在待测溶液中，磺基水杨酸是无色的，Fe^{3+} 的浓度很低，可认为无色，所以溶液的颜色源于配合物。因此，测定溶液的吸光度，就能求出有色配合物的组成和稳定常数。分光光度法测定配合物的组成和稳定常数，常采取浓比递变策略，即在保持金属离子浓度（c_M）和配体浓度（c_L）总和不变的前提下，通过改变两种溶液的配比，配制系列混合溶液，测定其溶液的吸光度（A）；以吸光度对配体的摩尔分数作图，得出曲线（图12-1），曲线上两条切线的延长线相交于 B，B 点的横坐标可用于推导出配合物（ML_n）的配位体数（方程12-1）：

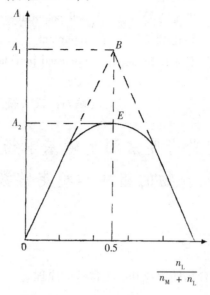

图12-1 浓比递变法测定配合物的组成

$$n = \frac{c_L}{c_M} = \frac{x_L}{x_M} = \frac{x_L}{1-x_L} \tag{12-1}$$

本实验中，由于测得配合物的组成比 $n=1$，所以该配合物可表示为 $[FeL]^+$。

假设 Fe^{3+} 和磺基水杨酸反应全部生成 $[FeL]^+$，在理想的图形中应该存在 B 点，它有最大吸光度值 A_1，而在实验曲线上，E 点的最大吸光度值 A_2 小于 A_1，A_1 和 A_2 的差值即可推出配合物的解离度和稳定常数。

配合物的解离度 α 应为：

$$\alpha = \frac{A_1 - A_2}{A_1}$$

$$M + L \rightleftharpoons ML$$

平衡浓度： $c\alpha \quad c\alpha \quad c(1-\alpha)$

所以，$K_{稳} = \dfrac{[ML]}{[M] \cdot [L]} = \dfrac{1-\alpha}{c \cdot \alpha^2}$

式中，c 为 B 点时金属离子 M 的浓度；$K_{稳}$ 为配合物 $[ML]$ 的稳定常数。

【实验仪器与药品】

分光光度计，容量瓶（100ml×2），烧杯（25ml×11），吸量管（10ml×2），pH 试纸。

H_2SO_4（浓），NaOH（6mol/L），磺基水杨酸（0.01mol/L），$NH_4Fe(SO_4)_2$ 溶液（0.01mol/L，在 pH=2 的 H_2SO_4 中）。

【实验步骤】

1. 配制 0.001mol/L $NH_4Fe(SO_4)_2$ 和 0.001mol/L 磺基水杨酸溶液各 100ml。

用吸量管移取 0.01mol/L $NH_4Fe(SO_4)_2$ 和 0.01mol/L 磺基水杨酸溶液各 10.00ml，分别置于两只 100ml 的容量瓶中，加纯化水稀释至刻度，注意要保证 pH 均为 2（在接近刻度时，检查 pH 值，若 pH 值偏离 2，可以滴加浓 H_2SO_4 或 6mol/L NaOH 进行调整）。

2. 用吸量管按表 12-1 中列出的体积，分别移取 0.001mol/L $NH_4Fe(SO_4)_2$ 和 0.001mol/L 磺基水杨酸溶液，置于 11 只 25ml 烧杯中，混匀。

3. 在 500nm 波长下，以纯化水为空白，测定上述 11 个溶液的吸光度。

【实验记录及数据处理】

1. 实验记录

表 12-1 实验记录

混合液编号	1	2	3	4	5	6	7	8	9	10	11
$V[NH_4Fe(SO_4)_2]$ (ml)	0	1.00	2.00	3.00	4.00	5.00	6.00	7.00	8.00	9.00	10.00
V（磺基水杨酸）(ml)	10.00	9.00	8.00	7.00	6.00	5.00	4.00	3.00	2.00	1.00	0
$\dfrac{n_L}{n_M+n_L}$											
A											

2. 数据处理

以吸光度（A）对摩尔分数 $\left(\dfrac{n_L}{n_M+n_L}\right)$ 作图，从图中即可求出配合物的组成和稳定常数，其中 $c = \dfrac{0.001 \times 5.00}{10.00}$。

Experiment 12　Determining the Composition and Stability Constant for Iron（Ⅲ）-Sulfosalicylic Acid Coordination Compound by Spectrophotometry

Preview

1. The usage of spectrophotometer.

2. The principle of determining the composition and stability constant for coordination compound by spectrophotometry.

Questions

1. Why can the formation of coordination compound be affected by acidity of solution?

2. Is the pH of each solution different in the experiment? If it is, how does the pH affect the experimental result?

3. What is the prerequisite for determining the composition and stability constant of coordination compound by spectrophotometry?

Principle

Sulfosalicylic acid (HOOC—  , simply "L" for short) reacts with Fe^{3+} to form stable coordination compound. The color and composition of coordination compounds vary with the pH values of solution. Coordination compound is purple $[FeL(H_2O)_4]^+$ when the pH is less than 4, red $[FeL_2(H_2O)_2]^-$ at the pH between 4 and 8, yellow $[FeL_3]^{3-}$ at the pH around 10, respectively. In this experiment, the composition and stability constant of purple coordination compound at the pH between 2 and 3 will be determined.

According to Lambert-Beer's law, when a beam of monochromatic light goes through colored solution, the absorbance of solution is directly proportional to the concentration of light-absorbing substance:

$$A = \varepsilon \, l \, c$$

where A is the absorbance of solution; c is the concentration of light-absorbing substance (mol/L); l is the thickness of solution (cm); ε is the molar absorption coefficient. When monochromatic light, solvent and temperature. are fixed, it is a characteristic constant of substance.

In the to-be-determined solution, since sulfosalicylic acid is colorless and Fe^{3+} can be considered to be colorless at low concentration, so the color of solution originates from coordination compound. Therefore, determining the absorbance of solution can calculate the composition and stability constant of colored coordination compound, which often takes the strategy of continuous concentration ratio variation by spectrophotometry, i. e., on the premise of keeping the sum of metal ion concentration (c_M) and ligand concentration (c_L) constant, a series of mixed solutions are prepared by changing ratios of two solutions, and absorbance of each solution is determined. The curve can be gotten by plotting the absorbance values vs. the molar fractions of ligand (as shown Fig. 12 – 1). The abscissa of intersection point B, which is obtained by prolonging two tangent lines, can be used to deduce the number of ligand of coordination compound (MLn) by (Eq. 12 – 1).

$$n = \frac{c_L}{c_M} = \frac{x_L}{x_M} = \frac{x_L}{1 - x_L} \qquad (12-1)$$

In this experiment, since the determined composition ratio n is equal to 1, so the coordination compound can be expressed as $[FeL]^+$.

If sulfosalicylic acid reacts with Fe^{3+} to form $[FeL]$ completely, there should be a point B in the curve, which possesses the maximum absorbance A_1. However, the maximum absorbance A_2 corresponding to point E in the experimental curve is less than A_1. The difference between A_1 and A_2 can be used to infer the dissociation degree and stability constant of coordination com-

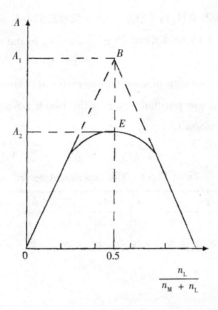

Fig. 12-1 Plot of continuous concentration ratio variation

pound.

The dissociation degree α of coordination compound is written as:

$$\alpha = \frac{A_1 - A_2}{A_1}$$

$$\text{M} + \text{L} \rightleftharpoons \text{ML}$$

Equilibrium concentration: $c\alpha \quad c\alpha \quad c(1-\alpha)$

$$K_s = \frac{[\text{ML}]}{[\text{M}] \cdot [\text{L}]} = \frac{1-\alpha}{c \cdot \alpha^2}$$

Where c is the concentration of metal ion M at point B, K_s is stability constant of coordination compound.

Instruments and chemicals

spectrophotometer, volumetric flask (100ml × 2), beaker (25ml × 11), mohr pipets (10ml ×2), pH test paper.

H_2SO_4 (concentrated), NaOH (6mol/L), sulfosalicylic acid (0.01mol/L), $NH_4Fe(SO_4)_2$ (0.01mol/L, in H_2SO_4 at pH = 2).

Procedures

1. Prepare 100ml of 0.001mol/L $NH_4Fe(SO_4)_2$ and 100ml of 0.001mol/L sulfosalicylic acid solution.

Transfer 10.00ml of 0.01mol/L $NH_4Fe(SO_4)_2$ and 10.00ml of 0.01mol/L sulfosalicylic acid accurately with mohr pipets, into two 100ml volumetric flasks, respectively, and dilute to scale mark with purified water and keep pH = 2 (Check the pH of solution when the liquid surface is close to scale mark, and drop concentrated H_2SO_4 or 6mol/L NaOH to adjust pH if necessary).

2. Transfer 0.001mol/L $NH_4Fe(SO_4)_2$ and 0.001mol/L sulfosalicylic acid accurately with mohr pipets into each of 11 beakers of 25ml, according to the volumes listed in Table 12-1, and mix them.

3. Set the detection wavelength of spectrophotometer at 500nm, then measure the each absorbance of 11 solutions and use purified water as the blank reference.

Records and Data processing

1. Data record

Table 12-1 The experiment record

No.	1	2	3	4	5	6	7	8	9	10	11
$V(NH_4Fe(SO_4)_2)$ (ml)	0	1.00	2.00	3.00	4.00	5.00	6.00	7.00	8.00	9.00	10.00
V(sulfosalicylic acid)(ml)	10.00	9.00	8.00	7.00	6.00	5.00	4.00	3.00	2.00	1.00	0
$\frac{n_L}{n_M + n_L}$											
A											

2. Data processing

Plot the absorbance values (A) vs. the mole fractions ($\frac{n_L}{n_M + n_L}$) for a series of solutions. From this curve, the composition and stability constant of the iron (Ⅲ)-sulfosalicylic acid coordination compound can be calculated, where $c = \frac{0.001 \times 5.00}{10.00}$.

Vocabulary

spectrophotometer	分光光度计
spectrophotometry	分光光度法
stability constant	稳定常数
sulfosalicylic acid	磺基水杨酸
vary with	随…而变化
Lambert-Beer's law	朗伯-比尔定律
a beam of monochromatic light	一束单色光
absorbance	吸光度
goes through colored solution	通过有色溶液
be directly proportional to	与…呈正比
molar absorption coefficient	摩尔吸光系数
intersection	交点
prolong tangent line	延长切线
dissociation degree	解离度
abscissa	横坐标
the difference value	差值

mole fraction	摩尔分数
volumetric flask	锥形瓶
mohr pipets	吸量管
concentrated H_2SO_4	浓硫酸
data processing	数据处理
blank reference	空白参比

（编写：刘迎春，英文核审：夏丹丹）

实验十三　配合物的生成和性质

【预习内容】

1. 配离子在水溶液中具有高稳定性的原因。
2. 配位平衡与其他平衡之间的关系。
3. 配合物的一些应用。

【思考题】

1. NH_3 和 NH_4^+ 都能做配体吗？为什么？
2. 哪些因素能影响配位平衡？
3. AgCl 和 AgBr 都能溶于 KCN 溶液吗？为什么？
4. 为什么能用草酸洗掉衣服上的铁锈？
5. 可用哪些类型的反应将 $[Fe(SCN)_n]^{3-n}$ 的红色褪去？

【实验原理】

1. 大多数金属离子都能与配体反应，生成配离子（或中性配合物）。金属离子、配体与配离子之间存在动态平衡，它被称作配位平衡。例如 $[Cu(NH_3)_4]^{2+}$，其配位平衡可表示为：

$$Cu^{2+} + 4NH_3 \rightleftharpoons [Cu(NH_3)_4]^{2+}$$

依据平衡移动原理，增加金属离子或配体的浓度，能促使配离子生成；相反，减少金属离子或配体浓度，则能促使配离子解离。

2. 在配合物的水溶液中，加入沉淀剂，会使配位平衡向配离子的解离方向移动。如：

$$[Ag(NH_3)_2]^+ + I^- \rightleftharpoons AgI\downarrow + 2NH_3$$

该反应的平衡常数为：

$$K = \frac{[NH_3]^2}{[Ag(NH_3)_2^+][I^-]} \cdot \frac{[Ag^+]}{[Ag^+]} = \frac{1}{K_{sp} \cdot K_{稳}} = \frac{1}{1.56 \times 10^{-16} \times 1.7 \times 10^7} = 3.77 \times 10^8$$

从 K 值看，正向反应很容易进行。

相反，向含有难溶盐的系统中加入配位剂，会使配位平衡向生成配离子的方向移动。如：

$$AgCl + 2NH_3 \rightleftharpoons [Ag(NH_3)_2]^+ + Cl^-$$

该反应平衡常数为：

$$K = \frac{[Ag(NH_3)_2^+][Cl^-]}{[NH_3]^2} \cdot \frac{[Ag^+]}{[Ag^+]} = K_{sp} \cdot K_{稳} = 1.56 \times 10^{-10} \times 1.7 \times 10^7 = 2.65 \times 10^{-3}$$

从 K 值看，虽然正向反应不是特别容易进行，但增大氨水浓度，可以促使上述平衡向右移动的。

由此可见，沉淀剂和配位剂与同一种金属离子反应的结果，主要取决于难溶盐的溶度积 K_{sp} 和配离子的稳定常数 $K_{稳}$。

3. 如果配体是弱碱根或弱酸根（如：NH_3、CN^-、$C_2O_4^{2-}$），向溶液中加入强酸，由于配体易与 H^+ 结合，促使配离子解离，这种现象称为酸效应，如：

$$[Cu(NH_3)_4]^{2+} + 4H^+ \rightleftharpoons Cu^{2+} + 4NH_4^+$$

溶液的酸度不仅影响配体，还会影响能够发生水解的金属离子。升高溶液的 pH 值，会促使配离子解离，这种现象称为水解效应。如：

$$[FeF_6]^{3-} + 3OH^- \rightleftharpoons Fe(OH)_3\downarrow + 6F^-$$

4. 配位平衡和氧化-还原平衡之间也存在相互作用，所以配离子的生成会改变电极电势值。如：

$$\varphi^\ominus([Co(NH_3)_6]^{3+}/[Co(NH_3)_6]^{2+}) = 0.10V, \quad \varphi^\ominus(Co^{3+}/Co^{2+}) = 1.84V$$

5. 在含有两种配体的溶液中，根据软硬酸碱理论，金属离子将更容易与硬度相似、浓度较大的配体结合。如：在 SCN^- 和较大浓度 F^- 共存的溶液中，Fe^{3+} 更易与 F^- 结合。

【实验仪器与药品】

离心机。

HCl（1mol/L），$NH_3 \cdot H_2O$（2mol/L、6mol/L），KI（0.1mol/L），KBr（0.1mol/L），$K_4[Fe(CN)_6]$（0.1mol/L），$K_3[Fe(CN)_6]$（0.1mol/L），NaCl（0.1mol/L），Na_2S（0.1mol/L），$Na_2S_2O_3$（1mol/L），EDTA-2Na（0.1mol/L），NH_4SCN（0.1mol/L，饱和），$(NH_4)_2C_2O_4$（饱和），$AgNO_3$（0.1mol/L），$CuSO_4$（0.1mol/L），$HgCl_2$（0.1mol/L），$FeCl_3$（0.1mol/L），Ni^{2+} 液，Fe^{3+} 和 Co^{2+} 混合液，I_2 液，丁二酮肟（1%），乙醇（95%），丙酮，NaF（s），锌粉（s）。

【实验内容】

1. 简单离子与配离子的区别 向含有 2 滴 0.1mol/L NH_4SCN 的两支试管中，分别加入 2 滴 0.1mol/L $FeCl_3$ 和 0.1mol/L $K_3[Fe(CN)_6]$，有何变化？解释现象。

2. 配离子稳定性的比较

（1）向含有 2 滴 0.1mol/L $FeCl_3$ 的试管中，加入数滴 0.1mol/L NH_4SCN，注意观察，然后逐滴加入 $(NH_4)_2C_2O_4$ 饱和溶液，观察溶液颜色的变化，写出反应方程式，并比较两种配离子的稳定性。

（2）在含有 10 滴 0.1mol/L $AgNO_3$ 的试管中，加入 10 滴 0.1mol/L NaCl，离心分离，除去上层清液，然后按次序做下列实验：

① 滴加 6mol/L 氨水，充分振荡试管至沉淀刚好溶解。
② 加 10 滴 0.1mol/L KBr，有何沉淀生成？
③ 除去上层清液，用 1mol/L $Na_2S_2O_3$ 代替 6mol/L 氨水，重复实验①。
④ 用 0.1mol/L KI 代替 0.1mol/L KBr，重复实验②，注意观察。
写出以上各反应的方程式，并根据实验现象比较下列方面：
① $[Ag(NH_3)_2]^+$ 和 $[Ag(S_2O_3)_2]^{3-}$ 的稳定常数。
② AgCl、AgBr 及 AgI 的溶度积。

（3）向有 0.5ml 碘液中，逐滴加入 0.1mol/L $K_4[Fe(CN)_6]$ 并充分振荡，有何现象？写出反应式。

结合 Fe^{3+} 可将 I^- 氧化成 I_2 的实验结果，比较 $\varphi^\ominus([Fe(CN)_6]^{3-}/[Fe(CN)_6]^{4-})$ 与 $\varphi^\ominus(Fe^{3+}/Fe^{2+})$ 的数值，并比较 $[Fe(CN)_6]^{3-}$ 和 $[Fe(CN)_6]^{4-}$ 稳定常数的大小。

3. 配位平衡的移动　在含有 5ml 的 0.1mol/L $CuSO_4$ 的小烧杯中加入 6mol/L 氨水，直至最初生成的 $Cu_2(OH)_2SO_4$ 沉淀刚好溶解。然后加入 6ml 的 95% 的乙醇，晶体将析出。过滤晶体，并用少量乙醇洗涤，观察晶体的颜色。写出反应式。

将少许的 $[Cu(NH_3)_4]SO_4$ 晶体，溶于 4ml 2mol/L 氨水中，得到 $[Cu(NH_3)_4]^{2+}$ 溶液。根据下述要求，设计使该配离子解离的实验方案，并写出反应式。
（1）用酸-碱反应破坏 $[Cu(NH_3)_4]^{2+}$。
（2）用沉淀反应破坏 $[Cu(NH_3)_4]^{2+}$。
（3）用氧化-还原反应破坏 $[Cu(NH_3)_4]^{2+}$。
提示：　　　$[Cu(NH_3)_4]^{2+} + 2e = Cu + 4NH_3$　　　$\varphi^\ominus = -0.02V$
　　　　　　$[Zn(NH_3)_4]^{2+} + 2e = Zn + 4NH_3$　　　$\varphi^\ominus = -1.02V$
（4）用配位反应破坏 $[Cu(NH_3)_4]^{2+}$。

4. 配合物的某些应用
（1）某些离子的鉴定　Ni^{2+} 与丁二酮肟反应生成鲜红色螯合物沉淀，用作 Ni^{2+} 的鉴定。

$$Ni^{2+} + 2\begin{array}{c}CH_3-C=NOH\\CH_3-C=NOH\end{array} \longrightarrow \begin{array}{c}CH_3-C=N\quad N=C-CH_3\\\quad\quad\quad Ni\quad\quad\quad\\CH_3-C=N\quad N=C-CH_3\end{array} \downarrow + 2H^+$$

在试管中，加入 2 滴 Ni^{2+} 试液、2 滴 2mol/L 氨水及 1 滴 1% 丁二酮肟，生成鲜红色沉淀。

（2）掩蔽干扰离子　在定性分析中，可以利用生成配合物的方法掩蔽干扰离子。例如，Co^{2+} 和 SCN^- 生成蓝色易溶于有机溶剂的 $[Co(SCN)_4]^{2-}$，可用于 Co^{2+} 的鉴定。Fe^{3+} 和 SCN^- 反应生成血红色的 $[Fe(SCN)_6]^{3-n}$。如果有 Fe^{3+} 和 Co^{2+} 共存时，血红色的 $[Fe(SCN)_6]^{3-n}$ 将干扰 Co^{2+} 的鉴定；由于 Fe^{3+} 与 F^- 反应生成无色的 $[FeF_6]^{3-}$，而且 $[FeF_6]^{3-}$ 比 $[Fe(SCN)_n]^{3-n}$ 更稳定，所以可用于"掩蔽" Fe^{3+}。

在含有 2 滴 Fe^{3+} 和 Co^{2+} 的混合试液的试管中，加入 8~10 滴饱和的 NH_4SCN 溶液，有何现象？加入 NaF 固体，并振荡试管，有何现象？最后加入 6 滴丙酮，记录现象。

（3）硬水软化　向盛有 50ml 自来水的烧杯中，加入 3~5 滴的 0.1mol/L EDTA-2Na，取另一只盛有 50ml 自来水的烧杯作对照，加热两只烧杯，煮沸 10 分钟后，未加入 EDTA-2Na 的烧杯中出现白色的悬浮物，而另外一只烧杯澄清，解释现象。

Experiment 13　Formation and Properties of Coordination Compounds

Preview

1. The reason for high stability of complex ions in aqueous solution.
2. The relationships between coordination equilibrium and other equilibriums.
3. Applications of coordination compounds.

Questions

1. Can both NH_3 and NH_4^+ be served as ligands? Why?
2. Which factors affect the coordination equilibrium?
3. Can both AgCl and AgBr be dissolved in KCN solution? Why?
4. Why can rust adhered on clothes be washed using oxalic acid?
5. Which types of reactions can be used to make $[Fe(SCN)_n]^{3-n}$ faded?

Principles

1. Most metal ions can react with ligands to form complex ions (or neutral coordination compounds). A dynamic equilibrium is established between the complex ion and metal ion, ligand(s), which is called coordination equilibrium. For $[Cu(NH_3)_4]^{2+}$, the equilibrium can be shown as

$$Cu^{2+} + 4NH_3 \rightleftharpoons [Cu(NH_3)_4]^{2+}$$

By the principle of equilibrium shift, increasing the concentration of metal ion or ligand(s) drives the equilibrium to form complex ion. Conversely, decreasing them will favor the dissociation of complex ion.

2. Involving precipitant in aqueous solution of coordination compound drives complexation equilibrium to the side of dissociation of complex ion. For example,

$$[Ag(NH_3)_2]^+ + I^- \rightleftharpoons AgI\downarrow + 2NH_3$$

The equilibrium constant for the reaction can be written as：

$$K = \frac{[NH_3]^2}{[Ag(NH_3)_2^+][I^-]} \cdot \frac{[Ag^+]}{[Ag^+]} = \frac{1}{K_{sp} \cdot K_s} = \frac{1}{1.56 \times 10^{-16} \times 1.7 \times 10^7} = 3.77 \times 10^8$$

From the value of K for the reaction, the dissociation of $[Ag(NH_3)_2]^+$ is much easy. Conversely, involving complexant in the system with insoluble salt dirves coordination equilibrium to the side of formation of complex ion. For example,

$$AgCl + 2NH_3 \rightleftharpoons [Ag(NH_3)_2]^+ + Cl^-$$

The equilibrium constant can be written as:

$$K = \frac{[Ag(NH_3)_2^+][Cl^-]}{[NH_3]^2} \cdot \frac{[Ag^+]}{[Ag^+]} = K_{sp} \cdot K_s = 1.56 \times 10^{-10} \times 1.7 \times 10^7 = 2.65 \times 10^{-3}$$

Although the formation of $[Ag(NH_3)_2]^+$ is less easy according to the value of K for the reaction, increasing the concentration of ammonia solution still drives the equilibrium to right.

It is concluded that the competition result of bonding to identical metal ion between precipitant and complexant mainly depends on solubility product K_{sp} for insoluble salt and stability constant K_s for complex ion.

3. If ligand is a weak base or the anion of a weak acid (for example: NH_3, CN^-, $C_2O_4^{2-}$), involving strong acid to the solution will make complex ion dissociated by the combination of ligand with H^+, which is called acid effect. For example,

$$[Cu(NH_3)_4]^{2+} + 4H^+ \rightleftharpoons Cu^{2+} + 4NH_4^+$$

The acidity of solution does not only affect ligand but also metal ion, which may be hydrolyzed. Increasing the pH of solution will make complex ion dissociated, which is called hydrolysis effect. For example,

$$[FeF_6]^{3-} + 3OH^- \rightleftharpoons Fe(OH)_3 \downarrow + 6F^-$$

4. There is interaction between coordination equilibrium and oxidation-reduction equilibrium. Thus, electrode potential can be altered by the formation of complex ion. For example,

$$\varphi^\ominus([Co(NH_3)_6]^{3+}/[Co(NH_3)_6]^{2+}) = 0.10V, \varphi^\ominus(Co^{3+}/Co^{2+}) = 1.84V$$

5. In a solution with two ligands, the metal ion will bond better to the ligand, which has similar hardness as it and bigger concentration, according to the soft-hard acid-base theory. For example, Fe^{3+} bond better to F^- than to SCN^- in a solution with SCN^- and concentrated F^-.

Instrument and Chemicals

centrifuger.

HCl (1mol/L), $NH_3 \cdot H_2O$ (2mol/L, 6mol/L), KI (0.1mol/L), KBr (0.1mol/L), $K_4[Fe(CN)_6]$ (0.1mol/L), $K_3[Fe(CN)_6]$ (0.1mol/L), NaCl (0.1mol/L), Na_2S (0.1mol/L), $Na_2S_2O_3$ (1mol/L), EDTA-2Na (0.1mol/L), NH_4SCN (0.1mol/L, saturated), $(NH_4)_2C_2O_4$ (saturated), $AgNO_3$ (0.1mol/L), $CuSO_4$ (0.1mol/L), $HgCl_2$ (0.1mol/L), $FeCl_3$ (0.1mol/L), Ni^{2+} solution, mixed solution of Fe^{3+} and Co^{2+}, iodine solution, dimethylglyoxime (1%), ethanol (95%), acetone, NaF (s), zinc powder (s).

Procedures

1. Difference between simple ion and complex ion

Add 2 drops of 0.1mol/L $FeCl_3$ and 2 drops of 0.1mol/L $K_3[Fe(CN)_6]$ into each two test tubes with 2 drops of 0.1mol/L NH_4SCN. What happens? Explain your observation.

2. Comparison of complex ions stability

(1) Add several drops of 0.1mol/L NH_4SCN into a test tube with 2 drops of 0.1mol/L $FeCl_3$. Note your observation. Then add $(NH_4)_2C_2O_4$ saturated solution dropwise, observe col-

or change of the solution, write out equations for this process and compare stability of two complex ions.

(2) Add 10 drops of 0.1mol/L NaCl into a test tube with 10 drops of 0.1mol/L $AgNO_3$, centrifugalize it and remove supernate, then do the tests in turn as follows:

①Add 6mol/L ammonia solution and shake well until precipitate has just been dissolved completely.

② Add 10 drops of 0.1mol/L KBr, what precipitate forms?

③Remove the supernate, and repeat test ① with 1mol/L $Na_2S_2O_3$ instead of 6mol/L aqueous ammonia.

④ Repeat test ② with 0.1mol/L KI instead of 0.1mol/L KBr. Note your observation.

Write out equations for those processes and compare the following aspects on the basis of your observations:

① Stability constants for $[Ag(NH_3)_2]^+$ and $[Ag(S_2O_3)_2]^{3-}$.

② Solubility products for AgCl, AgBr and AgI.

(3) Add 0.1mol/L $K_4[Fe(CN)_6]$ dropwise to 0.5ml of iodine solution and shake well. What happens? Write out the chemical equation.

Compare the values of $\varphi^\ominus([Fe(CN)_6]^{3-}/[Fe(CN)_6]^{4-})$ and $\varphi^\ominus(Fe^{3+}/Fe^{2+})$ according to the experimental result that Fe^{3+} can oxidize I^- into I_2, then compare the magnitude of stability constants for $[Fe(CN)_6]^{3-}$ and $[Fe(CN)_6]^{4-}$.

3. The shift of coordination equilibrium

Add 6mol/L aqueous ammonia in a small beaker with 5ml of 0.1mol/L $CuSO_4$ until initial $Cu_2(OH)_2SO_4$ has just been dissolved. Then add 6ml of 95% ethanol, and crystal of $[Cu(NH_3)_4]SO_4$ will precipitate. Filter the crystal and wash it with a small amount of ethanol. Observe the color of crystal and write out the reaction equations.

Dissolve a small portion of $[Cu(NH_3)_4]SO_4$ crystal in 4ml of 2mol/L $NH_3 \cdot H_2O$ to get $[Cu(NH_3)_4]^{2+}$ solution. Design tests to dissociate $[Cu(NH_3)_4]^{2+}$ according to the following requirements, and write out equations.

(1) Involve acid-base reaction in destroying $[Cu(NH_3)_4]^{2+}$.

(2) Involve precipitation reaction in destroying $[Cu(NH_3)_4]^{2+}$.

(3) Involve oxidation-reduction reaction in destroying $[Cu(NH_3)_4]^{2+}$.

Hints:

$$[Cu(NH_3)_4]^{2+} + 2e = Cu + 4NH_3 \qquad \varphi^\ominus = -0.02V$$
$$[Zn(NH_3)_4]^{2+} + 2e = Zn + 4NH_3 \qquad \varphi^\ominus = -1.02V$$

(4) Involve new ligand (or chelant) in destroying $[Cu(NH_3)_4]^{2+}$.

4. Applications of coordination compounds

(1) Identification of certain ions

Ni^{2+} reacts with dimethylglyoxime to form a bright red precipitate of chelate compound, which can be used for identification of Ni^{2+}.

$$Ni^{2+} + 2 \begin{matrix} CH_3-C=NOH \\ | \\ CH_3-C=NOH \end{matrix} \longrightarrow \begin{matrix} CH_3-C=N \\ | \\ CH_3-C=N \end{matrix} \underset{O\cdots H\cdots O}{\overset{O\cdots H\cdots O}{Ni}} \begin{matrix} N=C-CH_3 \\ | \\ N=C-CH_3 \end{matrix} \downarrow + 2H^+$$

Add 2 drops of Ni^{2+} solution, 2 drops of 2mol/L $NH_3 \cdot H_2O$ and 1 drop of 1% dimethylglyoxime into a test tube to yield a bright red precipitate.

(2) Masking of interference ions

In qualitative analysis, the method of forming coordination compound can be used to mask interference ion. For example, Co^{2+} reacts with SCN^- to form blue $[Co(SCN)_4]^{2-}$, which is soluble well in organic solvents. This reaction can be used to identify Co^{2+}. In addition, Fe^{3+} can also react with SCN^- to form blood red $[Fe(SCN)_n]^{3-n}$. If there is a mixture containing Fe^{3+} and Co^{2+}, the blood red $[Fe(SCN)_n]^{3-n}$ will interfere the identification of Co^{2+}. The reaction between Fe^{3+} and F^- can produce colorless $[FeF_6]^{3-}$ and $[FeF_6]^{3-}$ is more stable than $[Fe(SCN)_n]^{3-n}$. So F^- can be used to mask Fe^{3+}.

Add 8—10 drops of saturated NH_4SCN solution in a test tube with 2 drops of a mixture containing Fe^{3+} and Co^{2+}, what happens? Add solid NaF and shake well, what happens? At last, add 6 drops of acetone into the test tube, and note your observation.

(3) The softening of hard water

Add 3—5 drops of 0.1mol/L EDTA-2Na to a beaker with 50ml of tap water. There is another beaker with 50ml of tap water as a blank. Then heat the two beakers. White suspended solid appears in the beaker without EDTA-2Na after boiling for 10 minutes, while the other still keeps clean. Explain the difference in these two beakers, including equations if necessary.

Vocabulary

coordination compound, complex	配合物
coordination ion, complex ion	配离子
coordination agent, complexant	配位剂
chemical equilibrium	化学平衡
complexation	配合作用，配位
equilibrium shift	平衡移动
aqueous solution	水溶液
stability	稳定性
be served as	用作
be dissolved in	溶于
oxalic acid	草酸
dynamic equilibrium	动态平衡
precipitant	沉淀剂
precipitation	沉淀

dissociation	解离
insoluble salt	难溶盐
solubility product	溶度积
hydrolysis effect	水解效应
acid effect	酸效应
acidity	酸度
oxidation-reduction	氧化-还原
electrode potential	电极电势
soft-hard acid-base theory	软硬酸碱理论
concentrated	浓的
saturated solution	饱和溶液
centrifugalize	离心分离
supernate	上清液
precipitate（vt., vi.）	沉淀
oxidation-reduction	氧化还原
chelant	螯合剂
dimethylglyoxime	丁二酮肟
interference ion	干扰离子
filter（vt., vi.）	过滤

（实验编写：刘迎春，英文核审：夏丹丹）

实验十四　某些无机物分子或基团的空间构型

【预习内容】

1. 杂化轨道理论。
2. 价层电子对互斥理论。

【实验原理】

1. 杂化轨道理论要点

（1）原子轨道的杂化，只有在形成分子的过程中才会发生，而孤立的原子不可能进行杂化。原子轨道杂化只有在同一原子中，并且能量相近的原子轨道才能发生杂化。

（2）杂化轨道的数目与参与杂化的原子轨道数目相等。

（3）杂化轨道的成键能力比原来轨道成键能力增强。

（4）杂化轨道分为等性杂化和不等性杂化两种。

（5）杂化轨道类型与分子空间构型关系如表 14-1 所示。

表14-1 杂化轨道类型与分子空间构型关系

杂化类型	sp	sp^2	sp^3	dsp^2	sp^3d	sp^3d^2
空间构型	直线	平面（正）三角形	（正）四面体	平面（正）四边形	三角双锥	（正）八面体

2. 价层电子对互斥理论要点

（1）在一个共价型的AB_n分子或离子中，如果中心原子 A 的价电子层中 d 电子数为 0、5 或 10 时，则其空间构型完全由中心原子 A 的价电子层中的电子对数决定。价层电子对数等于 A 原子的价层电子数与 B 原子提供的电子数总和的一半。如果是离子，应该加上或减去离子电荷所需要的电子数。

（2）价层电子对的位置倾向于尽可能远离，使彼此间斥力最小，价层电子对数与理想空间构型关系如表 14-2 所示。

表14-2 价电子对数与理想几何构型关系

价层电子对数	2	3	4	5	6
理想空间构型	直线	平面（正）三角形	（正）四面体	三角双锥	（正）八面体

（3）中心原子 A 周围的孤电子对数等于价层电子对数与成键电子对数之差。

（4）在相同角度中（一般选 90°），电子对之间斥力大小的顺序为：

孤对电子 - 孤对电子 > 孤对电子 - 成键电子 > 成键电子 - 成键电子

（5）在 AB_n 中，若 A 与 B 之间通过双键或三键结合时，则按单键处理。

【实验材料】

塑料球、塑料棒。

【实验内容】

下面内容均用模型组装。

1. 根据杂化轨道理论判断分子或基团的空间构型（表 14-3）。

表14-3 实验记录

种类	$BeCl_2$	BCl_3	$[Zn(NH_3)_4]^{2+}$	CCl_4	NH_3	H_2O	$[FeF_6]^{3-}$
杂化类型							
键角							
分子的空间构型							

2. 根据价层电子对互斥理论判断一些非过渡元素共价化合物分子的空间构型。

（1）价层电子对互斥理论中的理想空间结构（表 14-4）

表14-4 实验记录

价层电子对数	价层电子对的几何排布	排布形式	角度
2	∶—A—∶	直线型	180°
3			
4			

续表

价层电子对数	价层电子对的几何排布	排布形式	角度
5			
6			

A 表示中心原子；∶表示电子对

（2）四氟化硫 SF_4（表 14-5）

表 14-5 实验记录

价层电子对数	成键电子对数	孤对电子对数	可能存在的构型（画图）		最稳定结构（画图并文字描述）
			a	b	
90°孤对-孤对排斥作用数					
90°孤对-键对排斥作用数					
90°键对-键对排斥作用数					

（3）三氟化氯 ClF_3 按照上述形式填写。
（4）四氟化氙 XeF_4 按照上述形式填写。
（5）二氟化氙 XeF_2 按照上述形式填写。
（6）排出下列各物质构型（表 14-6）

表 14-6 实验记录

实例	价层电子对数	成键电子对数	孤对电子对数	分子形状（文字描述）
$SnCl_2$				
NF_3				
PCl_5				
AsO_4^{3-}				
SF_6				
IF_5				
NH_4^+				
I_3^-				
SO_4^{2-}				
NO_2^-				
IO_3^-				
H_2O				

Experiment 14 Steric Configuration of Certain Inorganic Molecules or Groups

Preview

1. Hybrid orbital theory.

2. Valence-shell electron-pair repulsion (VSEPR) theory.

Principles

1. Main points of hybrid orbital theory

(1) The hybrid of atomic orbitals occurs during molecular formation process, which doesn't exist in the isolated atom. The atomic orbitals can only hybridize when they are in the same atom and have close energy level.

(2) The numbers of hybrid orbitals is equal to those of the hybridized atoms.

(3) The bonding capability is strengthened after hybrid of the atomic orbitals.

(4) There are two kinds of hybrid orbitals: equal hybrid and unequal hybrid.

(5) Different types of hybrid orbitals and corresponding steric configurations of molecules are shown in Table 14 – 1:

Table 14 – 1 Different types of hybrid orbitals and corresponding steric configurations of molecules

Type of hybrid orbitals	sp	sp^2	sp^3	dsp^2	sp^3d	sp^3d^2
Steric configuration	Linear	Trigonal planar	Tetrahedral	Square planar	Trigonal bipyramidal	Octahedral

2. Main points of valence-shell electron-pair repulsion theory

(1) If numbers of d electrons in central atoms are 0, 5 or 10, the shape of covalent AB_n molecules or ions can be determined from the numbers of electron pairs in valence shell of central atom A. The numbers of electron pairs in valence shell equal to half of the sum of (the numbers of electrons in valence shell of central atom A + the total numbers of electrons from B atoms ± the charge numbers of AB_n ions). Here, if AB_n is cationic, we should subtract the numbers of charge; AB_n is anionic, we should plus the numbers of charge.

(2) To keep that the repulsions among them are minimized, the electron pairs in valence shell should be apart from each other as far as possible. The numbers of electrons in valence shell and the corresponding ideal geometric configurations are shown in Table 14 – 2:

Table 14 – 2 The relationship between the numbers of electrons in valence shell and ideal geometric configuration

Number of electrons in valence shell	2	3	4	5	6
Ideal geometric configuration	Linear	Trigonal planar	Tetrahedral	Trigonal bipyramidal	Octahedral

(3) The numbers of lone pair of electrons around central atom A equal to the numbers of electron pairs in valence shell subtract the numbers of bonding pair of electrons.

(4) Among the same angles (choosing 90° generally), the order of repulsions is:

A lone pair of electrons-a lone pair of electrons > a lone pair of electrons-a bonding pair of electrons > a bonding pair of electrons-a bonding pair of electrons

(5) It should be treated as single bond if there is a double bond or a triple bond in a AB_n-type molecular or group.

Materials

Plastic balls, plastic sticks.

Procedures

The following steps are all assembled with plastic model.

1. Determine the steric configuration of the following molecules or groups by hybrid orbital theory; fill in Table 14 – 3.

Table 14 – 3 Experiment record

Type	$BeCl_2$	BCl_3	$[Zn(NH_3)_4]^{2+}$	CCl_4	NH_3	H_2O	$[FeF_6]^{3-}$
Kinds of hybrid							
Band angle							
Steric configuration of molecular							

2. Determine the steric configuration of some covalent compounds including non-transition elements by valence-shell electron-pair repulsion theory.

(1) Fill in Table 14 – 4 according to valence-shell electron-pair repulsion theory

Table 14 – 4 Experiment record

Numbers of electron pairs in valence shell	Geometrical of electron pairs in valence shell	Shape	Angle
2	：—A—：	straight line	180°
3			
4			
5			
6			

A represents central atom, ： represents electron pair

(2) Sulfur tetrafluoride SF_4

Table 14 – 5 Experiment record

Numbers of electron pairs in valence shell	Numbers of bonding pair of electrons	Numbers of lone pair of electrons	Possible Shape (draw a picture)		The most stable configuration (draw a picture and descript in words)
			a	b	
Numbers of 90° repulsion between a lone pair of electrons and a lone pair of electrons					
Numbers of 90° repulsion between a lone pair of electrons and a bonding pair of electrons					
Numbers of 90° repulsion between a bonding pair of electrons and a bonding pair of electrons					

(3) Chlorine trifluoride ClF_3, fill in according to the above form.

(4) Xenon tetrafluoride XeF_4, fill in according to the above form.

(5) Xenon difluoride XeF_2, fill in according to the above form.

(6) Point out the configuration of following matters.

Table 14-6 Experiment record

Experiment	Numbers of electron pairs in valence shell	Numbers of bonding pair of electrons	Numbers of lone pair of electrons	Molecular shape (descript in words)
$SnCl_2$				
NF_3				
PCl_5				
AsO_4^{3-}				
SF_6				
IF_5				
NH_4^+				
I_3^-				
SO_4^{2-}				
NO_2^-				
IO_3^-				
H_2O				

Vocabulary

| orbital | 电子轨道 | see-saw | 跷跷板 |
| steric configurations | 空间构型 | | |

（编写：兰阳，英文核审：夏丹丹）

第四章 元素化学实验

Chapter 4　Elements Chemistry Experiments

实验十五　p区元素

一、卤素、氧、硫

【预习内容】

1. 卤素单质及其有关化合物的化学性质。
2. H_2O_2 的化学性质。
3. H_2S 的化学性质及各类型硫化物的生成条件。
4. 不同氧化态硫的化合物的主要性质。

【思考题】

1. 淀粉碘化钾试纸遇高浓度氯气，或较长时间与氯气接触，会观察到什么现象，原因何在？
2. 哪些物质既能作氧化剂又能作还原剂？H_2O_2 被氧化和被还原的产物是什么？
3. H_2S、Na_2S、Na_2SO_3 的溶液放置久了，会发生什么变化？如何判断变化情况？
4. $Na_2S_2O_3$ 溶液和 $AgNO_3$ 溶液反应，试剂的用量不同产物有何不同？

【仪器与试剂】

试管，水浴。

固体：I_2，KI，NaCl，KBr，KI，MnO_2，Na_2SO_3，$K_2S_2O_8$。

酸：H_2SO_4（2mol/L，浓），HCl（2mol/L）。

碱：NaOH（2mol/L），$NH_3 \cdot H_2O$（浓）。

盐：KBr（0.1mol/L），KI（0.1mol/L），Pb$(NO_3)_2$（0.1mol/L），$KMnO_4$（0.01mol/L），$K_2Cr_2O_7$（0.1mol/L），$Na_2S_2O_3$（0.1mol/L），$AgNO_3$（0.1mol/L），$MnSO_4$（0.1mol/L），$ZnSO_4$（0.1mol/L），Pb$(NO_3)_2$（0.1mol/L），$CuSO_4$（0.1mol/L）。

其他：溴水，CCl_4，氯水，饱和 H_2S 水溶液，碘液，Pb$(Ac)_2$ 试纸，H_2O_2（3%），淀粉溶液，乙醚。

【操作步骤】

1. 卤素单质

（1）溴和碘的溶解性

① 在试管中加入 5 滴溴水，再加入 5 滴 CCl_4，充分振荡后，观察 CCl_4 层的颜色。比较溴在水和 CCl_4 中的溶解性。

② 取一小粒碘于试管中，加 1ml 纯化水，振荡试管，观察溶液的颜色，再加入少量 KI 晶体，振荡后观察溶液的颜色变化，比较碘在水中和在 KI 溶液中的溶解性。然后在试管中继续加入 10 滴 CCl_4，充分振荡，观察水层和 CCl_4 层的颜色，比较碘在水和 CCl_4 中的溶解性。

（2）卤素的歧化反应

① 在试管中加入 5 滴溴水，滴加数滴 2mol/L NaOH，振荡后观察现象。

② 用碘液代替溴水进行实验，写出反应方程式。

（3）卤素单质的氧化性比较

① 在盛有 2 滴 0.1mol/L KBr 溶液的试管中，加入 2 滴氯水，再加入 5 滴 CCl_4 并振荡，观察 CCl_4 层的颜色，试比较氯和溴的氧化性。

② 在盛有 2 滴 0.1mol/L KI 溶液的试管中，加入 2 滴溴水，再加入 5 滴 CCl_4 并振荡，观察 CCl_4 层的颜色，试比较溴和碘的氧化性。

③ 在盛有 1 滴 0.1mol/L KI 溶液的试管中，加入 2 滴氯水，并加入 5 滴 CCl_4，观察 CCl_4 层的颜色。再继续滴加过量的氯水至 CCl_4 层颜色消失，写出有关的反应方程式。

④ 取碘液 2 滴，滴加饱和 H_2S 水溶液，至碘液的颜色消失，写出反应方程式。

根据以上实验结果，比较卤素单质的氧化性。

2. 卤素离子的还原性比较

（1）往盛有少量 NaCl 固体的干燥试管中加入 2~3 滴浓 H_2SO_4，有何现象？用玻璃棒蘸一些浓 $NH_3·H_2O$，移近试管口，以检验气体产物 HCl。

（2）往盛有少量 KBr 固体的干燥试管中加入 2~3 滴浓 H_2SO_4，有何现象？用润湿的淀粉-KI 试纸在管口检验气体产物。

（3）往盛有少量 KI 固体的干燥试管中加入 2~3 滴浓 H_2SO_4，有何现象？用润湿的 $Pb(Ac)_2$ 试纸在管口检验气体产物 H_2S。

综合上述实验，比较氯、溴、碘离子的还原性。

3. 氯酸盐和碘酸盐的氧化性

（1）氯酸钾的氧化性　在试管中加入 1ml 饱和 $KClO_3$ 溶液和数滴 0.1mol/L KI 溶液，把得到的混合液分成 2 份，一份用数滴 2mol/L H_2SO_4 酸化，另一份留作比较，振荡试管，观察溶液有何变化。比较氯酸盐在中性溶液和酸性溶液中氧化性的强弱。

（2）碘酸钾的氧化性　在试管中加入 1ml 饱和 KIO_3 溶液和数滴 2mol/L H_2SO_4，再加入 0.1mol/L KI 溶液，振荡试管，观察现象，写出反应方程式。

4. 过氧化氢

（1）过氧化氢的氧化性

① 取 2 滴 0.1mol/L $Pb(NO_3)_2$ 溶液，加入 4 滴饱和 H_2S 水溶液，观察沉淀颜色，

然后加入3% H_2O_2溶液，直至沉淀颜色转变为白色，写出反应方程式。

②取5滴0.1mol/L KI溶液，用3滴2mol/L H_2SO_4酸化，再加入2滴3% H_2O_2溶液，并加入2滴淀粉溶液检验产物，写出反应方程式。

(2) 过氧化氢的还原性　在试管中加入5滴0.01mol/L $KMnO_4$溶液，用3滴2mol/L H_2SO_4酸化后，滴入数滴3% H_2O_2溶液至溶液紫红色褪去，写出反应方程式。

(3) 过氧化氢的催化分解　在盛有10滴3% H_2O_2溶液的试管中，加入少量MnO_2固体，管口用带火星的火柴杆检验O_2的生成。

(4) 过氧化氢的鉴定　在试管中加入0.5ml纯化水，再加入2滴0.1mol/L $K_2Cr_2O_7$和2滴2mol/L H_2SO_4，并加入10滴乙醚。然后加入2滴3% H_2O_2溶液，振荡试管，观察乙醚层的蓝色。CrO_5不稳定，会慢慢分解为Cr^{3+}，乙醚层蓝色逐渐褪去。反应方程式如下：

$$Cr_2O_7^{2-} + 4H_2O_2 + 2H^+ = 2CrO_5 + 5H_2O$$

$$4CrO_5 + 12H^+ = 4Cr^{3+} + 7O_2\uparrow + 6H_2O$$

5. 硫化氢和金属硫化物

(1) 硫化氢的还原性　在一支试管中加入2滴0.01mol/L $KMnO_4$溶液，用5滴6mol/L H_2SO_4酸化，加入数滴饱和H_2S水溶液；在另一支试管中加入1滴0.1mol/L $K_2Cr_2O_7$溶液，用5滴6mol/L H_2SO_4酸化，再加入数滴饱和H_2S水溶液。分别观察现象，并写出反应方程式。

(2) 难溶硫化物　在4支试管中分别加入5滴0.1mol/L $ZnSO_4$、$Pb(NO_3)_2$、$CuSO_4$、$AgNO_3$溶液，然后各滴加饱和H_2S水溶液，观察各试管中有无沉淀生成。若无沉淀，继续加入2mol/L 氨水至溶液呈碱性，观察各试管中沉淀的颜色。

6. 亚硫酸盐的性质

(1) 与酸反应　取少量Na_2SO_3晶体于试管中，加入2mol/L H_2SO_4，将品红滴在滤纸上，在试管口检验所产生的气体，保留H_2SO_3溶液供下面实验使用。

(2) 氧化性　取实验(1)所得的H_2SO_3溶液，加入饱和H_2S水溶液，观察硫的析出。

(3) 还原性　取5滴0.01mol/L $KMnO_4$溶液，加入实验(1)所得的H_2SO_3溶液，观察现象，写出反应方程式。

7. 硫代硫酸盐的性质

(1) 与酸反应　在试管中加入0.5ml 0.1mol/L $Na_2S_2O_3$溶液，并加入0.5ml 2mol/L HCl，观察现象，并检验产生的气体。

(2) 还原性　在试管中加入2滴碘液和1滴淀粉溶液，滴加0.1mol/L $Na_2S_2O_3$溶液，至溶液蓝色褪去，写出反应方程式。

在试管中加入2滴0.1mol/L $Na_2S_2O_3$溶液，加入数滴氯水，并检验反应中生成的SO_4^{2-}，写出反应方程式。

(3) $S_2O_3^{2-}$的鉴定　在试管中加入2滴0.1mol/L $Na_2S_2O_3$溶液，并逐滴加入0.1mol/L $AgNO_3$溶液，先产生$Ag_2S_2O_3$白色沉淀，沉淀很快转变为黄色、棕色，最后变为棕黑色，反应方程式如下：

$$Ag_2S_2O_3 + H_2O = 2H^+ + SO_4^{2-} + Ag_2S\downarrow$$

8. 过二硫酸盐的氧化性 把 2ml 2mol/L H_2SO_4、2ml 纯化水和 2 滴 0.1mol/L $MnSO_4$ 溶液混合均匀，分成 2 份。在第一份中加 1 滴 0.1mol/L $AgNO_3$ 溶液和少量 $K_2S_2O_8$ 固体（催化剂 Ag^+ 可加快反应），水浴加热，观察溶液颜色有何变化；在第二份中只加少量 $K_2S_2O_8$ 固体，水浴加热，观察溶液颜色各有何变化。反应方程式如下：

$$5S_2O_8^{2-} + 2Mn^{2+} + 8H_2O = 10SO_4^{2-} + 2MnO_4^- + 16H^+$$

【注意事项】

1. 氯气是有强烈刺激性气味的剧毒气体，进行有关氯气的试验，必须在通风橱中操作。

2. 溴蒸气对呼吸道有很强的刺激作用，用量应尽可能少，并在装有吸收装置或在通风橱内进行。溴水有较强的腐蚀性，使用溴水时，应避免与皮肤接触。

3. 氯酸钾是强氧化剂，绝不允许将氯酸钾与硫、磷混在一起。氯酸钾易分解，不宜研磨、烘干。进行有关氯酸钾的试验，剩下的应放入专用的回收瓶内。

4. 硫化氢及二氧化硫都是有毒气体。特别是硫化氢气体极毒，吸入微量即发生头痛、眩晕，制备和使用时要在通风橱中操作。

5. 过氧化物是氧化剂，对皮肤有腐蚀性，使用时应注意。

Experiment 15　p Block Elements

I. Halogen, oxygen, sulfur

Preview

1. Chemical properties of halogen and related compounds.

2. Chemical properties of H_2O_2.

3. Chemical properties of H_2S and formation conditions of sulfides.

4. The properties of oxidation states of sulfur in different compounds.

Questions

1. What will occur when amylum-KI test paper exposes to high concentration of chlorine or contacts with chlorine completely for quite some time? Explain your answer.

2. What kind of substances can be both oxidant and reductant? What is the oxidation or reduction product of H_2O_2?

3. Explain why H_2S, Na_2S, Na_2SO_3 solution can not be stored for a long time. Prove your answer.

4. What will occur when different dosage of $Na_2S_2O_3$ solution reacts with $AgNO_3$ solution?

Instruments and Chemicals

test tubes, water bath.

solids: I_2, KI, NaCl, KBr, KI, MnO_2, Na_2SO_3, $K_2S_2O_8$.

acids: H_2SO_4(2mol/L, concentrated), HCl(2mol/L).

bases: NaOH (2mol/L), NH$_3 \cdot$H$_2$O(concentrated).

salts: KBr(0.1mol/L), KI(0.1mol/L), Pb(NO$_3$)$_2$(0.1mol/L), KMnO$_4$(0.01mol/L), K$_2$Cr$_2$O$_7$(0.1mol/L), Na$_2$S$_2$O$_3$(0.1mol/L), AgNO$_3$(0.1mol/L), MnSO$_4$(0.1mol/L), ZnSO$_4$(0.1mol/L), Pb(NO$_3$)$_2$(0.1mol/L), CuSO$_4$(0.1mol/L).

others: aqueous of bromine, CCl$_4$, aqueous solution of chlorine, Pb(Ac)$_2$ test paper, saturated aqueous H$_2$S, iodine solution, H$_2$O$_2$(3%), starch solution, ether.

Procedures

1. Elemental halogen

(1) Solubility of bromine and iodine

① Add five drops of CCl$_4$ into a test tube with 5 drops of aqueous bromine, then note the color of CCl$_4$ layer after shaking sufficiently. Compare the solubility of Br$_2$ in water and CCl$_4$.

② Place an iodine granule into a test tube. Add 1ml purified water, then note the color of the iodine solution after shaking. Add a small amount of KI crystal into the same test tube, and note the color change of the solution after shaking sufficiently. Compare the solubility of I$_2$ in water and KI solution. Then 10 drops of CCl$_4$ is added into the same tube, shake sufficiently, and note the color of aqueous layer and CCl$_4$ layer, respectively. Compare the solubility of I$_2$ in water, aqueous KI and CCl$_4$.

(2) Disproportionation of halogen

① Add 5 drops of aqueous bromine and several drops of 2mol/L NaOH solution into a test tube. Note your observation after shaking.

② Repeat test ①, using aqueous iodine instead of aqueous bromine. Note your observation and write out the equation for this process.

(3) Comparison the oxidizing activity of elemental halogen

① Add 3 drops of aqueous chlorine into a test tube with 2 drops of 0.1mol/L KBr solution, then add 5 drops of CCl$_4$. Note the color of the CCl$_4$ layer after shaking. Explain your observation.

② Add 2 drops of aqueous bromine into a test tube with 2 drops of 0.1mol/L KI solution, then add 5 drops of CCl$_4$. Note the color of the CCl$_4$ layer after shaking. Explain your observation.

③ Add 1 drop of aqueous chlorine into a test tube with 2 drops of 0.1mol/L KI solution, then add 5 drops of CCl$_4$. Note the color of the CCl$_4$ layer after shaking. Then add excess aqueous chlorine into the same test tube until the color of the CCl$_4$ layer disappears. Write out the relevant reaction equations.

④ Add several drops of saturated aqueous H$_2$S into a test tube with 2 drops of iodine solution until the iodine solution becomes colorless. Write out the balanced equation for this process.

Summarize the experimental results above, and compare the oxidizing activity of elemental halogen.

2. Comparison of reducing activity of the halogen ions

(1) Add 2 or 3 drops of concentrated H_2SO_4 into a dry test tube with a small amount of NaCl. Note down your observation. Put a glass rod dipped with aqueous ammonia near the top of the tube to detect the gas released (HCl). Explain your observation.

(2) Add 2 or 3 drops of concentrated H_2SO_4 into a dry test tube with several KBr pellets. Note down your observation. Put a piece of wetting amylum-KI test paper near the top of the tube to detect the gas released.

(3) Add 2 or 3 drops of concentrated H_2SO_4 into a dry test tube with several KI pellets. Note down your observation. Put a piece of wetting Pb(Ac)$_2$ test paper near the top of the tube to detect the gas released (H_2S).

Summarize the experimental results above, and compare the reducing activity of elemental halogen.

3. Oxidizing activity of chlorates and iodates

(1) Oxidizing activity of potassium chlorate

Add 1ml saturated $KClO_3$ solution and several drops of 0.1mol/L KI solution into a test tube, then the mixture solution above is divided into two portions, one portion is acidified with several drops of 2mol/L H_2SO_4, the other is reserved for comparison. Note your observation after shaking. Compare the oxidizing activity of the chlorates in different medium, e.g. in neutral or acidic solution.

(2) Oxidizing activity of potassium iodate

Add 1ml saturated KIO_3 solution and several drops of 2mol/L H_2SO_4 solution into a test tube, then add 0.1mol/L KI solution into the mixture solution above, shake the tube, observe the phenomena, and write out the reaction equation.

4. Peroxide

(1) Oxidizing activity of H_2O_2

① Add 4 drops of saturated H_2S solution into a test tube with 2 drops of 0.1mol/L Pb(NO$_3$)$_2$ solution. Note the color of the precipitate, then a few drops of 3% H_2O_2 solution is added into the test tube until the precipitate turns white. Write out the reaction equation.

② Take 5 drops of 0.1mol/L KI solution acidified with 3 drops of 2mol/L H_2SO_4 solution, then add 2 drops of 3% H_2O_2 solution into the test tube, 2 drops of starch solution is added to detect the product. Write out the reaction equation.

(2) Reducing activity of H_2O_2

Add 5 drops of 0.01mol/L $KMnO_4$ solution acidified with 3 drops of 2mol/L H_2SO_4 solution into a test tube, then add 3% H_2O_2 solution into the solution above until violet disappears. Write out the reaction equation.

(3) The decompositions of H_2O_2 under catalyst

Add a small amount of MnO_2(s) into a test tube with 10 drops of 3% H_2O_2 solution, then put a sparking match near the outlet of the tube to detect O_2.

(4) Identification of H_2O_2

Add 0.5ml purified water, 2 drops of 0.1mol/L $K_2Cr_2O_7$ solution and 2 drops of 2mol/L H_2SO_4 solution into a test tube. 10 drops of ether are added into the solution, then add 2 drops of 3% H_2O_2 solution, shake the tube, and observe the color (blue) of ether layer. CrO_5 is unstable, so it will decompose to Cr^{3+} slowly, the color of the ether layer disappears gradually. The equations for this process are as follows

$$Cr_2O_7^{2-} + 4H_2O_2 + 2H^+ = 2CrO_5 + 5H_2O$$
$$4CrO_5 + 12H^+ = 4Cr^{3+} + 7O_2\uparrow + 6H_2O$$

5. Hydrogen sulfide and metal sulfides

(1) Reducing activity of hydrogen sulfide

Add 2 drops of 0.01mol/L $KMnO_4$ solution into a test tube, acidified with 5 drops of 6mol/L H_2SO_4 solution, then several drops of saturated H_2S solution is added into the tube. Add a drop of 0.1mol/L $K_2Cr_2O_7$ solution into another test tube, acidified with 5 drops of 6mol/L H_2SO_4 solution, then several drops of saturated H_2S solution is added into the tube. Note your observation, and write out the reaction equations, respectively.

(2) Insoluble metal sulfides

Add 5 drops of 0.1mol/L $ZnSO_4$, $Pb(NO_3)_2$, $CuSO_4$, $AgNO_3$ solution into four test tubes, respectively, then several drops of saturated H_2S solution is added into each tube. Observe whether the precipitate is formed in each tube. If there is no precipitate in some tubes, continue to add 2mol/L aqueous ammonia until the solution is alkaline. Observe the color of the precipitates in each tube.

6. Properties of sulfite

(1) Reaction with acid

Add 2mol/L H_2SO_4 solution into a test tube with a small amount of Na_2SO_3 crystal, cover the outlet of the tube with a filter paper impregnated with magenta to detect the gas released. Keep the left H_2SO_3 solution for the next experiment.

(2) Oxidizing activity

Add several drops of saturated aqueous H_2S into a test tube with the H_2SO_3 solution prepared in test (1). Observe the emergence of sulfur.

(3) Reducing activity

Take 5 drops of 0.01mol/L $KMnO_4$ solution, then add the H_2SO_3 solution prepared in test (1). Note your observation, and write out the reaction equation.

7. Properties of thiosulfate

(1) Reaction with acid

Add 0.5ml 2mol/L HCl solution into a test tube with 0.5ml 0.1mol/L $Na_2S_2O_3$ solution, Note your observation and detect the gas released.

(2) Reducing activity

Add 0.1mol/L $Na_2S_2O_3$ solution into a test tube with 2 drops of iodine solution and a drop of starch solution. Note the color change of the solution and write out the reaction equation.

Add several drops of aqueous chlorine into a test tube with 2 drops of 0.1mol/L $Na_2S_2O_3$ solution. How to confirm the formation of SO_4^{2-} ion? Write out the reaction equation.

(3) Identification of $S_2O_3^{2-}$

Add 2 drops of 0.1mol/L $Na_2S_2O_3$ solution into a test tube, and add 0.1mol/L $AgNO_3$ solution into the tube dropwise. First a white precipitate appears, then the precipitate promptly transforms into yellow, brown and finally turns brown black. The reaction equation is as follows

$$Ag_2S_2O_3 + H_2O = 2H^+ + SO_4^{2-} + Ag_2S \downarrow$$

8. Oxidizing activity of persulfate

Mix 2ml 2mol/L H_2SO_4, 2ml purified water and 2 drops of 0.1mol/L $MnSO_4$ solution. The mixture is divided into two portions. Add one drop of 0.1mol/L $AgNO_3$ solution and a small amount of $K_2S_2O_8$ solid into the first portion (Ag^+ catalyst can accelerate the reaction), while only add a small amount of $K_2S_2O_8$ solid into the second portion. Heat the tubes with water bath. Note the color change of the solution, respectively. The reaction equation is as follows.

$$5S_2O_8^{2-} + 2Mn^{2+} + 8H_2O = 10SO_4^{2-} + 2MnO_4^- + 16H^+$$

Notes

1. Chlorine is hazardous gas with offensive odor. Handle only in a fume hood.

2. Bromine vapor has a strong stimulating effect on the tract. Dosage should be minimized. Handle only with an absorbing device or in a fume hood. Aqueous bromine is strongly corrosive. Please avoid contacting with skin while operating.

3. Potassium chlorate is a strong oxidizing agent. Never allow the potassium chlorate and sulfur, phosphorus mixed together in order to avoid an explosion. Potassium chlorate is easy to decompose. Don't grind and dry it. For experiments carried out on potassium chlorate, the rest must be placed in a designed waste container.

4. Hydrogen sulfide and sulfur dioxide are both poisonous gases. Especially hydrogen sulfide gas is highly toxic. Inhaling trace gas can cause headache and dizziness. Handle in a fume hood.

5. Peroxide is an oxidizing agent and corrosive. Be careful to deal with it.

Vocabulary

halogen	卤素	chlorine	氯
bromine	溴	iodine	碘
potassium iodide	KI	potassium bromide	KBr
sodium chloride	NaCl	concentrated sulfuric acid	浓硫酸
hydrogen sulfide	H_2S	concentrated aqueous ammonia	浓氨水
carbon tetrachloride	CCl_4	hydrogen peroxide	H_2O_2
potassium chlorate	$KClO_3$	potassium iodate	KIO_3
sodium sulfite	Na_2SO_3	chromium peroxide	CrO_5

ether	乙醚	sulfur dioxide	SO_2
sodium hydroxide	NaOH	sulfuric acid	H_2SO_4
potassium persulfate	$K_2S_2O_8$		

二、氮族、锡、铅

【预习内容】

1. 硝酸的强氧化性。亚硝酸的不稳定性、氧化性和还原性。
2. 磷酸各种钙盐溶解性的不同。
3. 砷、锑、铋氧化物、氢氧化物的酸碱性，砷、锑、铋盐的氧化还原性。
4. 锡、铅氢氧化物的酸碱性。二价锡的还原性、四价铅的氧化性。

【思考题】

1. 在氧化还原反应中，为什么一般不用硝酸、盐酸作为反应的酸性介质？在哪种情况下可以用它们做酸性介质？
2. 如何鉴别 $NaNO_2$ 和 $NaNO_3$ 溶液？
3. 实验室配制 $SnCl_2$ 溶液时，为何既要加盐酸又要加锡粒？
4. 试验 $Pb(OH)_2$ 的碱性时，应选何种酸为宜？
5. 在用 $NaBiO_3$ 氧化 Mn^{2+} 的实验中，为什么要取少量 $MnSO_4$ 溶液？若 Mn^{2+} 过多会对实验有何影响？

【仪器与试剂】

试管，水浴。

固体：锌粒，As_2O_3（极毒），$NaBiO_3$，PbO_2。

酸：H_2SO_4（2mol/L，6mol/L），HCl（2mol/L，6mol/L，浓），HNO_3（2mol/L，6mol/L，浓）。

碱：NaOH（2mol/L，6mol/L），$NH_3 \cdot H_2O$（2mol/L）。

盐：$NaNO_2$（1mol/L），Na_3PO_4（0.1mol/L），Na_2HPO_4（0.1mol/L），NaH_2PO_4（0.1mol/L），$CaCl_2$（0.1mol/L），$SbCl_3$（0.1mol/L），$Bi(NO_3)_3$（0.1mol/L），KI（0.1mol/L），$K_2Cr_2O_7$（0.1mol/L），$MnSO_4$（0.1mol/L），$KMnO_4$（0.01mol/L），$SnCl_2$（0.1mol/L，0.5mol/L），$Pb(NO_3)_2$（0.1mol/L），$FeCl_3$（0.1mol/L），$HgCl_2$（0.1mol/L）。

其他：奈氏试剂，淀粉-KI试纸，淀粉溶液。

【操作步骤】

1. 硝酸的氧化性

（1）分别往3支各盛有一小粒锌粒的小试管中加入浓 HNO_3、6mol/L HNO_3、2mol/L HNO_3 各约2ml，比较各试管主要反应产物有何不同？

（2）待反应片刻后，把2mol/L HNO_3 与锌反应得到的溶液倒在另一支试管中，并慢慢加入6mol/L NaOH 溶液，直到最初生成的白色沉淀溶解后再滴加几滴，加热试管，用奈氏试剂湿润的试纸检验产生的气体，观察现象，写出反应方程式。

2. 亚硝酸的生成和性质

（1）亚硝酸的生成和分解　在试管中加入5滴1mol/L NaNO$_2$溶液，然后加入3滴6mol/L H$_2$SO$_4$，观察溶液的颜色和液面上方气体的颜色，写出相应的反应方程式。

（2）亚硝酸的氧化性　在试管中加入2滴0.1mol/L KI溶液，加入纯化水稀释至1ml，用2滴2mol/L H$_2$SO$_4$酸化，然后加入2滴1mol/L NaNO$_2$溶液，观察现象。再加入2滴淀粉溶液检验I$_2$的生成，写出反应方程式。

（3）亚硝酸的还原性　在试管中加入5滴0.01mol/L KMnO$_4$溶液，并加入2滴2mol/L H$_2$SO$_4$酸化，然后加入2滴1mol/L NaNO$_2$溶液，观察溶液的颜色变化，写出反应方程式。

3. 磷酸盐的性质　用pH试纸分别检验0.1mol/L Na$_3$PO$_4$、Na$_2$HPO$_4$、NaH$_2$PO$_4$溶液的pH值。

在3支试管中各加入3滴0.1mol/L Na$_3$PO$_4$、Na$_2$HPO$_4$和NaH$_2$PO$_4$溶液，然后再分别加入10滴0.1mol/L CaCl$_2$溶液，观察哪个有沉淀产生？向没有产生沉淀的溶液中滴加2mol/L NH$_3$·H$_2$O，观察有何变化？最后检验所有沉淀是否能溶于6mol/L HCl。比较Ca$_3$（PO$_4$）$_2$、CaHPO$_4$和Ca（H$_2$PO$_4$）$_2$的溶解度，说明它们之间相互转化的条件。

4. 砷、锑、铋的氧化物或氢氧化物

（1）As$_2$O$_3$的性质　在2支试管中各加入少量As$_2$O$_3$（极毒）。在一支试管中加入少量2mol/L NaOH，振荡，观察As$_2$O$_3$是否溶解［保留溶液供As（Ⅲ）的还原性实验使用］？在另一支试管中加入6mol/L HCl，振荡，观察As$_2$O$_3$溶解情况，必要时可微热溶液。根据实验结果说明As$_2$O$_3$的酸碱性。

（2）Sb（OH）$_3$的生成和性质　在2支试管中各加入3滴0.1mol/L SbCl$_3$溶液，再分别加入3滴2mol/L NaOH，观察白色Sb（OH）$_3$的生成。然后分别滴加2mol/L NaOH和2mol/L HCl，沉淀是否溶解？写出反应方程式。

（3）Bi（OH）$_3$的生成和性质　将0.1mol/L Bi（NO$_3$）$_3$溶液与2mol/L NaOH作用制得两份Bi（OH）$_3$，分别检验沉淀在2mol/L NaOH、2mol/L HCl中的溶解情况。对Bi（OH）$_3$的酸碱性作出结论。

5. 砷、铋的氧化还原性

（1）As（Ⅲ）的还原性和As（Ⅴ）的氧化性　在试管中加入3滴0.1mol/L KI溶液和1滴淀粉溶液，并加入Na$_3$AsO$_4$溶液［实验（1）保留］，观察现象，再滴加6mol/L HCl，观察I$_2$的析出。然后再滴加6mol/L NaOH，观察现象，写出反应方程式。

（2）Bi（Ⅴ）的氧化性　在试管中加入2滴0.1mol/L MnSO$_4$溶液，加入5滴6mol/L HNO$_3$酸化（能否用HCl酸化？），然后加入少量NaBiO$_3$固体，观察溶液的颜色变化，反应方程式如下（此反应常用来鉴定Mn^{2+}）：

$$2Mn^{2+} + 5NaBiO_3 + 14H^+ = 2MnO_4^- + 5Bi^{3+} + 5Na^+ + 7H_2O$$

6. 锡、铅氢氧化物的生成和性质

（1）Sn(OH)$_2$的生成和性质　在2支试管中各加入2滴0.5mol/L SnCl$_2$溶液，滴入2mol/L NaOH，观察Sn（OH）$_2$沉淀的生成。在Sn（OH）$_2$沉淀中分别加入6mol/L HCl

和 6mol/L NaOH，观察 Sn(OH)₂ 沉淀是否溶解。

(2) Pb(OH)₂ 的生成和性质　在 2 支试管中各加入 2 滴 0.1mol/L Pb(NO₃)₂ 溶液，分别滴入 2mol/L NaOH，观察 Pb(OH)₂ 沉淀的生成。分别试验它与稀酸与稀碱（适宜用哪种酸？）的反应。

7. 锡、铅的氧化还原性

(1) Sn(Ⅱ) 的还原性

① 在试管中加入 3 滴 0.1mol/L FeCl₃ 溶液和 2 滴 0.1mol/L KSCN 溶液，然后加入数滴 0.5mol/L SnCl₂ 溶液，至溶液血红色消失。写出反应方程式。

② 在试管中加入 1 滴 0.1mol/L HgCl₂ 溶液，逐滴加入 0.5mol/L SnCl₂ 溶液，观察生成的 Hg₂Cl₂ 白色沉淀。继续滴加过量的 SnCl₂ 溶液，振荡试管，放置片刻，观察沉淀的颜色变化，反应方程式如下（此反应可用来鉴定 Sn^{2+} 或 Hg^{2+}）：

$$2HgCl_2 + Sn^{2+} = Hg_2Cl_2 \downarrow + Sn^{4+} + 2Cl^-$$
$$Hg_2Cl_2 + Sn^{2+} = 2Hg \downarrow + Sn^{4+} + 2Cl^-$$

(2) Pb(Ⅳ) 的氧化性

① 在试管中加入少量 PbO₂ 固体，然后滴加 2 滴浓 HCl，观察现象，用润湿的淀粉 – KI 试纸检验 Cl₂ 的生成，并写出反应方程式。

② 在试管中加入少量 PbO₂ 固体，再加入 5 滴 6mol/L HNO₃ 和 2 滴 0.1mol/L MnSO₄ 溶液，搅拌，水浴加热，观察溶液的颜色变化，写出反应方程式。

【注意事项】

1. 亚硝酸及其盐有毒，注意勿入口内。

2. 砷、锑、铋化合物都有毒性，特别是砷的氧化物（俗称砒霜）是剧毒物质，要在教师指导下使用，取用量要少，切勿进入口内或与有伤口的地方接触。实验后一定要洗手。

II. The group VA elements, Tin and Lead

Preview

1. The strong oxidation of nitric acid. The instability, oxidation and reduction of nitrite.

2. The main properties of phosphate.

3. The acid and alkali strength of arsenic, antimony, bismuth oxides and hydroxides, oxidizing and reducing properties of arsenic and bismuth compounds.

4. The acid and alkali strength of tin, lead hydroxides. The reducing activity of Sn(Ⅱ) and the oxidizing activity of Pb(Ⅳ).

Questions

1. Why are nitric acid or hydrochloric acid not used as the acidic medium in the redox reaction generally? When can nitric acid and hydrochloric acid be used as acidic medium?

2. How to identify NaNO₂ and NaNO₃ solution?

3. Why is it necessary to add hydrochloric acid and tin pellets when SnCl₂ solution is prepared in laboratory?

4. What kind of acid should be used to test the alkaline of $Pb(OH)_2$?

5. Why does the concentration of $MnSO_4$ need to be controlled for the oxidation of Mn^{2+} to MnO_4^{2-} using $NaBiO_3$? What will happen if $MnSO_4$ is excessive?

Instruments and Chemicals

test tubes, water bath.

solid: Zn, As_2O_3 (highly toxic), $NaBiO_3$, PbO_2.

acids: H_2SO_4 (2mol/L, 6mol/L), HCl (2mol/L, 6mol/L, concentrated), HNO_3 (2mol/L, 6mol/L, concentrated).

bases: $NaOH$ (2mol/L, 6mol/L), $NH_3 \cdot H_2O$ (2mol/L).

salts: $NaNO_2$ (1mol/L), Na_3PO_4 (0.1mol/L), Na_2HPO_4 (0.1mol/L), NaH_2PO_4 (0.1mol/L), $CaCl_2$ (0.1mol/L), $SbCl_3$ (0.1mol/L), $Bi(NO_3)_3$ (0.1mol/L), KI (0.1mol/L), $K_2Cr_2O_7$ (0.1mol/L), $MnSO_4$ (0.1mol/L), $KMnO_4$ (0.01mol/L), $SnCl_2$ (0.1mol/L, 0.5mol/L), $Pb(NO_3)_2$ (0.1mol/L), $FeCl_3$ (0.1mol/L), $HgCl_2$ (0.1mol/L).

others: Nessler reagent, amylum-KI test paper, starch solution.

Procedures

1. Oxidizing activity of nitric acid

(1) Add about 2ml concentrated HNO_3 solution, 6mol/L HNO_3 solution and 2mol/L HNO_3 solution into three different tubes with a small pellets of zinc, respectively. Compare the main products in different test tubes.

(2) After a while pour the solution obtained by the reaction of zinc with 2mol/L HNO_3 into another test tube, and add 6mol/L $NaOH$ solution slowly until a white precipitate originally generated dissolves, then continue to add several drops of 6mol/L $NaOH$ solution, heat the tube, and detect the gas released by a test paper impregnated with Nessler reagent. Note your observation and write out the equation for the overall process.

2. Preparation and properties of nitrite salts

(1) Preparation and decomposition of nitrous acid

Add 5 drops of 1mol/L $NaNO_2$ solution and 3 drops of 6mol/L H_2SO_4 solution into a test tube. Observe the color of the solution and the gas releasing above the solution level and write out the equation for this process.

(2) Oxidizing activity of nitrous acid

Add 2 drops of 0.1mol/L KI solution into a test tube, diluted with purified water to 1ml and acidified with 2 drops of 2mol/L H_2SO_4 solution, then add 2 drops of 1mol/L $NaNO_2$ solution. Note your observation. Add two drops of starch solution to detect the formation of I_2. Write out the reaction equation.

(3) Reducing activity of nitrous acid

Add 5 drops of 0.01mol/L $KMnO_4$ solution into a test tube, acidified with 2 drops of 2mol/L H_2SO_4 solution, then 2 drops of 1mol/L $NaNO_2$ solution is added into the tube. Observe the color change of the solution and write out the reaction equation.

3. Properties of phosphates

Measure the pH of 0.10mol/L Na_3PO_4、Na_2HPO_4、NaH_2PO_4 solution with pH test paper, respectively.

Add 2 drops of 0.1mol/L Na_3PO_4, Na_2HPO_4, and NaH_2PO_4 solution into three different test tubes, then add ten drops of 0.1mol/L $CaCl_2$ solution into each tube. Observe whether the precipitate generates or not. Add several drops of 2mol/L aqueous ammonia into the solution without precipitate generated. Note your observation. Then test whether all the precipitates dissolve in 6mol/L HCl solution. Compare the solubility of $Ca_3(PO_4)_2$, $CaHPO_4$ and $Ca(H_2PO_4)_2$. Illustrate the conditions of their mutual transformation.

4. Oxide or hydroxide of arsenic, antimony, bismuth

(1) Properties of As_2O_3

Add a small amount of As_2O_3(highly toxic) into each of the two test tubes. Add several drops of 2mol/L NaOH solution into one tube. Observe whether As_2O_3 dissolves or not after shaking (reserve the solution for the reduction test of As (Ⅲ)). Add a small amount of 6mol/L HCl solution into the other one. Note your observation after shaking, gently heated if necessary. Illustrate the acidity and alkality of As_2O_3 according to the experimental results.

(2) Preparation and properties of $Sb(OH)_3$

Add 3 drops of 0.1mol/L $SbCl_3$ solution into two test tubes, respectively. Add 3 drops of 2mol/L NaOH solution into each tube. Observe the formation of the white $Sb(OH)_3$ precipitate, then add 2mol/L HCl solution and 2mol/L NaOH solution into each tube, respectively. Observe whether the precipitate dissolves or not and write out the reaction equation.

(3) Preparation and properties of $Bi(OH)_3$

Prepare two portions of $Bi(OH)_3$ precipitate with 0.1mol/L $Bi(NO_3)_3$ solution and 2mol/L NaOH solution. Test whether the precipitate dissolves in 2mol/L NaOH or 2mol/L HCl solution. Conclude the acidity and alkality of $Bi(OH)_3$.

5. Redox properties of arsenic, bismuth

(1) Reducing activity of As (Ⅲ) and oxidizing activity of As (Ⅴ)

Add 3 drops of 0.1mol/L KI solution and one drop of starch solution into a test tube, then add 5 drops of Na_3AsO_4 solution (obtained from experiment (1)). Note your observation. Then add 6mol/L HCl solution. Observe the formation of iodine. Finally add drops of 6mol/L NaOH solution. Note your observation and write out the reaction equation.

(2) Oxidizing activity of Bi (Ⅴ)

Add 2 drops of 0.1mol/L $MnSO_4$ solution into a test tube, acidified with 5 drops of 6mol/L HNO_3 solution (Can the solution be acidified with HCl solution, and why?), then add a small amount of $NaBiO_3$ solid into the tube with stirring. Observe the color change of the solution. The equation for this process is as follows(the identification of Mn^{2+})

$$2Mn^{2+} + 5NaBiO_3 + 14H^+ = 2MnO_4^- + 5Bi^{3+} + 5Na^+ + 7H_2O$$

6. Preparation and properties of hydroxides of tin, lead

(1) Preparation and properties of $Sn(OH)_2$

Add drops of 2mol/L NaOH solution into two different test tubes with 2 drops of 0.1mol/L $SnCl_2$ solution. Observe the formation of $Sn(OH)_2$ precipitate. 6mol/L HCl and 6mol/L NaOH solution is added into the $Sn(OH)_2$ precipitate, respectively. Observe whether $Sn(OH)_2$ is dissolved or not.

(2) Preparation and properties of $Pb(OH)_2$

Add drops of 2mol/L NaOH solution into two different test tubes with 2 drops of 0.1mol/L $Pb(NO_3)_2$ solution. Observe the formation of $Pb(OH)_2$ precipitate. Test the precipitate with dilute acids (which acid is suitable?) and dilute alkali, respectively. Note and explain the result.

7. Redox properties of tin, lead

(1) Reducing activity of Sn (Ⅱ)

①Add 3 drops of 0.1mol/L $FeCl_3$ solution and 2 drops of 0.1mol/L KSCN solution into a test tube, then add several drops of 0.5mol/L $SnCl_2$ solution into the tube until the blood red solution fades. Write out the reaction equation.

②Add one drop of 0.1mol/L $HgCl_2$ solution into a test tube, add 0.5mol/L $SnCl_2$ solution dropwise. Observe the white Hg_2Cl_2 precipitate. Continue to drop excessive $SnCl_2$ solution, shake the tube, and observe the color change of the precipitate after kept for a moment. The reaction equations are as follows (the identification of Sn^{2+} or Hg^{2+})

$$2HgCl_2 + Sn^{2+} = Hg_2Cl_2 \downarrow + Sn^{4+} + 2Cl^-$$
$$Hg_2Cl_2 + Sn^{2+} = 2Hg \downarrow + Sn^{4+} + 2Cl^-$$

(2) Oxidizing activity of Pb (Ⅳ)

①Add a small amount of PbO_2 solid into a test tube, then add 2 drops of concentrated HCl, observe the phenomena, detect Cl_2 with wet amylum-KI test paper and write out the reaction equation.

②Add a small amount of PbO_2 solid into a test tube with 5 drops of 6mol/L HNO_3 solution and 2 drops of 0.1mol/L $MnSO_4$ solution. Stir and heat the tube in water bath. Observe the color change of the solution and write out the reaction equation.

Notes

1. Nitrous acid and its salts are toxic. Make sure not to enter the mouth.

2. Arsenic, antimony, bismuth and their compounds are hazardous. Especially arsenic oxide commonly known as arsenic is highly toxic. It must be used under the supervising of your instructor. Dosage should be minimized, and make sure not to enter the mouth and contact with the wound. Always wash your hands after finished with your experiments.

Vocabulary

| nitrogen | 氮 | phosphorus | 磷 |
| arsenic | 砷 | antimony | 锑 |

bismuth	铋	tin	锡
lead	铅	zinc	锌
nitric acid	HNO_3	nitrite	HNO_2
sodium nitrite	$NaNO_2$	sodium phosphate	Na_3PO_4
sodium hydrogen phosphate	Na_2HPO_4	calcium phosphate	$Ca_3(PO_4)_2$
sodium dihydrogen phosphate	NaH_2PO_4	arsenic trioxide	As_2O_3
calcium hydrogen phosphate	$CaHPO_4$	antimony trichloride	$SbCl_3$
calcium dihydrogen phosphate	$Ca(H_2PO_4)_2$	antimony hydroxide	$Sb(OH)_3$
bismuth nitrate	$Bi(NO_3)_3$	bismuth hydroxide	$Bi(OH)_3$
sodium arsenite	Na_3AsO_4	manganese sulfate	$MnSO_4$
amylum-KI test pape	淀粉-KI 试纸	bismuth Sodium	$NaBiO_3$
tin dichloride	$SnCl_2$	lead nitrate	$Pb(NO_3)_2$
lead hydroxide	$Pb(OH)_2$	mercuric chloride	$HgCl_2$
lead dioxide	PbO_2	concentrated hydrochloric acid	浓 HCl

（编写：赵兵，英文核审：夏丹丹）

实验十六　d 区元素和 ds 区元素

一、d 区元素

【预习内容】

1. 铬常见的氧化态及它们存在的状态和颜色。
2. 锰常见的氧化态及它们存在的状态和颜色。
3. $Cr(OH)_3$ 和 $Mn(OH)_2$ 均是两性氢氧化物吗？
4. $Cr_2O_7^{2-}$ 与 CrO_4^{2-} 相互转化的条件是什么？它们形成相应盐的溶解性大小是多少？
5. $KMnO_4$ 的还原产物和介质有什么关系？

【思考题】

1. CrO_2^- 被氧化为 CrO_4^{2-} 的介质条件是什么，为什么？
2. 在酸性和碱性介质中，Cr(Ⅲ)、Cr(Ⅵ) 之间的转化反应如何发生？
3. 哪几种氧化剂可以将 Mn^{2+} 氧化成 MnO_4^-？实验时为什么 Mn^{2+} 不能过量？
4. 在制备 $Fe(OH)_2$ 的实验中，为什么纯化水和 NaOH 溶液须事先经过煮沸？
5. 为什么 Fe^{3+} 能把 I^- 氧化成 I_2，而 $[Fe(CN)_6]^{3-}$ 则不能？为什么 $[Fe(CN)_6]^{4-}$ 能把 I_2 还原为 I^-，而 Fe^{2+} 则不能？试从配合物的生成对电极电势的影响来解释。

【仪器与试剂】

试管，水浴。

固体：Na_2SO_3，$(NH_4)_2Fe(SO_4)_2$，KOH，$KClO_3$，MnO_2。

酸：H_2SO_4（2mol/L，6mol/L），HCl（2mol/L），HAc（2mol/L）。

碱：$NaOH$（2mol/L，6mol/L），$NH_3 \cdot H_2O$（2mol/L）。

盐：$CrCl_3$（0.1mol/L），Na_2S（0.5mol/L），K_2CrO_4（0.1mol/L），$AgNO_3$（0.1mol/L），$Pb(NO_3)_2$（0.1mol/L），$BaCl_2$（0.1mol/L），$K_2Cr_2O_7$（0.1mol/L），$MnSO_4$（0.1mol/L），$KMnO_4$（0.01mol/L），$FeCl_3$（0.1mol/L），KI（0.1mol/L），$K_4[Fe(CN)_6]$（0.1mol/L），$K_3[Fe(CN)_6]$（0.1mol/L）。

其他：饱和 NH_4Cl 溶液，氯水，淀粉溶液，碘液。

【操作步骤】

1. 铬的化合物

（1）Cr（Ⅲ）的化合物

① $Cr(OH)_3$ 的生成和性质：在 2 支试管中各加入 3 滴 0.1mol/L $CrCl_3$ 溶液，再分别滴加 1 滴 2mol/L $NaOH$，观察灰绿色 $Cr(OH)_3$ 沉淀的生成。分别用 2mol/L HCl 和 2mol/L $NaOH$ 检验 $Cr(OH)_3$ 是否具有两性。

② Cr（Ⅲ）的还原性：在①中得到的 $Cr(OH)_4^-$（即 $CrO_2^- + 2H_2O$）的溶液中加入数滴 3% H_2O_2 溶液，观察溶液的颜色变化，写出有关反应方程式。

③ Cr（Ⅲ）的水解：在试管中加入 5 滴 0.1mol/L $CrCl_3$ 溶液，检验其酸碱性，再加入 3 滴 0.5mol/L Na_2S 溶液，观察现象，并证明产物是 $Cr(OH)_3$ 而不是 Cr_2S_3。

（2）Cr（Ⅵ）的化合物

① $Cr_2O_7^{2-}$ 和 CrO_4^{2-} 的相互转化：在试管中加入 2 滴 0.1mol/L K_2CrO_4 溶液，滴入少量 2mol/L H_2SO_4，观察溶液的颜色变化。继续加入 2mol/L $NaOH$，观察溶液颜色又有何变化？写出反应方程式。

② 难溶性铬酸盐：在 3 支试管中分别加入 2 滴 0.1mol/L K_2CrO_4 溶液，再各加入 2 滴 0.1mol/L $AgNO_3$、$Pb(NO_3)_2$、$BaCl_2$ 溶液。观察产物的颜色，写出反应方程式。

以 $K_2Cr_2O_7$ 溶液代替 K_2CrO_4 溶液，作同样的实验，有什么现象？写出反应方程式。

③ Cr（Ⅵ）的氧化性：在 2 支试管中各加入 3 滴 0.1mol/L $K_2Cr_2O_7$ 溶液，分别用 3 滴 6mol/L H_2SO_4 酸化。在一支试管中加入少量 Na_2SO_3 晶体，在另一支试管中加入少量 $(NH_4)_2Fe(SO_4)_2$ 晶体，观察现象，并写出反应方程式。

④ Cr（Ⅵ）的检验：同 H_2O_2 的鉴定。

2. 锰的化合物

（1）Mn（Ⅱ）的化合物：在 4 支试管中分别加入 3 滴 0.1mol/L $MnSO_4$ 溶液，然后依次加入 2mol/L $NaOH$，观察 $Mn(OH)_2$ 沉淀的颜色。将 1 支试管在空气中振荡，观察沉淀的颜色变化。另外 3 支试管中分别加入 2mol/L HCl、2mol/L $NaOH$ 和饱和 NH_4Cl 溶液，观察沉淀是否溶解。

（2）Mn（Ⅳ）的化合物

① MnO_2 的生成：在试管中加入 3 滴 0.01mol/L $KMnO_4$ 溶液，再加 5 滴 0.1mol/L $MnSO_4$ 溶液，观察棕色沉淀的生成。

② MnO_2 的氧化性：在①中得到的 MnO_2 沉淀中加入数滴 6mol/L H_2SO_4 和少量 Na_2

SO_3 晶体，观察沉淀是否溶解？

(3) Mn(Ⅵ) 的化合物

① K_2MnO_4 的制备：在干燥试管中加入少量 MnO_2 和约等量的 $KClO_3$ 固体，再加入 2 粒固体 KOH，混匀后加热至熔融。冷却后加适量水使熔块溶解，得到深绿色的 K_2MnO_4 溶液，保留溶液做以下实验。

② K_2MnO_4 的歧化反应：取①中 K_2MnO_4 溶液 5 滴，加入 5 滴 2mol/L HAc，观察溶液的颜色变化和 MnO_2 的生成，写出反应方程式。

③ K_2MnO_4 的还原性：取①中 K_2MnO_4 溶液 2 滴，滴入氯水，观察溶液的颜色变化，写出反应方程式。

(4) Mn(Ⅶ) 的化合物：在 3 支试管中分别加入 2 滴 0.01mol/L $KMnO_4$ 溶液，再分别加入 2 滴 2mol/L H_2SO_4、水和 6mol/L NaOH，然后各加入少量 Na_2SO_3 晶体，观察各试管中所发生的现象，写出反应式，并说明 $KMnO_4$ 被还原的产物与介质的关系。

3. 铁的化合物

(1) $Fe(OH)_2$ 的生成和性质：在 1 支试管中加入 2ml 纯化水和 3 滴 6mol/L H_2SO_4，煮沸，然后溶入少量 $(NH_4)_2Fe(SO_4)_2$ 晶体。在另一试管中加入 3ml 6mol/L NaOH 溶液，煮沸，并迅速加入到 $(NH_4)_2Fe(SO_4)_2$ 溶液中（不要振荡），观察现象。静置片刻，观察沉淀的颜色变化。

用同样方法再制备一份 $Fe(OH)_2$，立即加入 2mol/L HCl，观察沉淀是否溶解？写出反应方程式。

(2) $Fe(OH)_3$ 的生成和性质：在盛有 1ml 0.1mol/L $FeCl_3$ 溶液的试管中，加入 2mol/L NaOH，观察沉淀的颜色。然后往试管中加入 0.5ml 2mol/L HCl，沉淀是否溶解？

(3) 铁的配合物

① 在一支试管中加入 2 滴 0.1mol/L $FeCl_3$ 溶液，再加入 2 滴 0.1mol/L KI 溶液和 1 滴淀粉溶液。在另一支试管中加入 2 滴 0.1mol/L $K_3[Fe(CN)_6]$ 溶液，再加入 2 滴 0.1mol/L KI 溶液和 1 滴淀粉溶液。各有何现象发生？

② 在试管中加入 2 滴 0.1mol/L $K_4[Fe(CN)_6]$ 溶液，并加入 2 滴碘液，然后加入少量 $(NH_4)_2Fe(SO_4)_2$ 晶体，观察蓝色沉淀或溶胶（滕氏蓝）的生成（此反应可用来鉴定 Fe^{2+}）。

③ 在试管中加入 2 滴 0.1mol/L $FeCl_3$ 溶液，再加入 1 滴 0.1mol/L $K_4[Fe(CN)_6]$ 溶液，观察深蓝色沉淀或溶胶（普鲁士蓝）的生成（此反应可用来鉴定 Fe^{3+}）。

【注意事项】

铬及其化合物均是有毒物质。特别是 Cr(Ⅵ) 毒性很大，Cr(Ⅵ) 化合物不仅对消化道和皮肤有强刺激性，而且有致癌作用；使用时取量要少，实验后废液倒入指定的废液回收容器内统一处理。

Experiment 16 d Block Elements and ds Block Elements

I. d Block Elements

Preview

1. Variable oxidation states and colors of chromium compounds.

2. Variable oxidation states and colors of manganese compounds.

3. Are $Cr(OH)_3$ and $Mn(OH)_2$ both amphoteric hydroxides?

4. What are the conditions for interconversion between $Cr_2O_7^{2-}$ and CrO_4^{2-}? Note the solubility difference between chromates and dichromate.

5. Reduction products of $KMnO_4$ are changing with medium.

Questions

1. In which medium can CrO_2^- be oxidized to CrO_4^{2-}, and why?

2. Note the interconversion between Cr(Ⅲ) and Cr(Ⅵ) in acidic and alkaline medium.

3. What kind of oxidants can oxidize Mn^{2+} to MnO_4^-? Why need the concentration of Mn^{2+} be controlled in this process?

4. Why need purified water and NaOH solution be pre-boiled, when you prepare $Fe(OH)_2$ precipitate?

5. Explain the following processes according to the effect of complex ion formation on electrode potential. Fe^{3+} ion can oxidize I^- to I_2 while $[Fe(CN)_6]^{3-}$ can't. $[Fe(CN)_6]^{4-}$ ion can reduce I_2 to I^- while Fe^{2+} can't.

Instrumentsand Chemicals

test tubes, water bath.

solid: Na_2SO_3, $(NH_4)_2Fe(SO_4)_2$, KOH, $KClO_3$, MnO_2.

acids: H_2SO_4 (2mol/L, 6mol/L), HCl (2mol/L), HAc (2mol/L).

bases: NaOH (2mol/L, 6mol/L), $NH_3 \cdot H_2O$ (2mol/L).

salts: $CrCl_3$ (0.1mol/L), Na_2S (0.5mol/L), K_2CrO_4 (0.1mol/L), $AgNO_3$ (0.1mol/L), $Pb(NO_3)_2$ (0.1mol/L), $BaCl_2$ (0.1mol/L), $K_2Cr_2O_7$ (0.1mol/L), $MnSO_4$ (0.1mol/L), $KMnO_4$ (0.01mol/L), $FeCl_3$ (0.1mol/L), KI (0.1mol/L), $K_4[Fe(CN)_6]$ (0.1mol/L), $K_3[Fe(CN)_6]$ (0.1mol/L).

others: saturated NH_4Cl solution, aqueous solution of chlorine, CCl_4, starch solution, iodine solution.

Procedures

1. Chromium compounds

(1) Compounds of Cr(Ⅲ)

① Preparation and properties of $Cr(OH)_3$

Add a drop of 2mol/L NaOH solution into two test tubes with 3 drops of 0.1mol/L $CrCl_3$

solution, respectively, and observe the grey green $Cr(OH)_3$ precipitate. Confirm the amphoteric property of $Cr(OH)_3$ by adding 2mol/L HCl and 2mol/L NaOH solution, respectively.

② Reducing activity of Cr(Ⅲ)

Add a small amount of 3% H_2O_2 solution into the $Cr(OH)_4^-$ ($CrO_2^- + 2H_2O$) solution prepared in test ①, then observe the color change of the solution, and write out the equation for this process.

③ Hydrolysis of Cr(Ⅲ)

Add 5 drops of 0.1mol/L $CrCl_3$ solution into a test tube, test the pH value of the solution, then add 3 drops of 0.5mol/L Na_2S solution. Note your observation and confirm that the product is $Cr(OH)_3$ not Cr_2S_3.

(2) Compounds of Cr(Ⅵ)

① Interconversion between $Cr_2O_7^{2-}$ and CrO_4^{2-}

Add 2 drops of 0.1mol/L K_2CrO_4 solution into a test tube, then add a small amount of 2mol/L H_2SO_4 solution, and observe the color change of the solution. If 2mol/L NaOH solution is added into the tube above, what will occur? Write out the equation for this process.

② Insoluble chromate

Add 2 drops of 0.1mol/L $AgNO_3$, $Pb(NO_3)_2$, $BaCl_2$ solution into each of the three tubes with 2 drops of 0.1mol/L K_2CrO_4 solution, then observe the color of the products. Write out the equations for processes above.

Repeat the previous test with 0.1mol/L $K_2Cr_2O_7$ solution instead of 0.1mol/L K_2CrO_4 solution. Note your observation and write out the chemical equations.

③ Oxidizability of Cr(Ⅵ)

Add 3 drops of 0.1mol/L $K_2Cr_2O_7$ solution into each of the two test tubes, with 3 drops of 6mol/L H_2SO_4 solution to acidify the solution. A small amount of Na_2SO_3 and $(NH_4)_2Fe(SO_4)_2$ crystal is added into the test tubes above respectively. Note your observation and write out the equations for this process.

④ The identification of Cr(Ⅵ)

Refer to the identification of H_2O_2.

2. Manganese compounds

(1) Compounds of Mn(Ⅱ)

Add 3 drops of 0.1mol/L $MnSO_4$ solution into four test tubes respectively, then add 2mol/L NaOH solution into each of the tubes successively. Observe the color of $Mn(OH)_2$ precipitate. Shake one test tube in the air. Note the color change of the precipitate. Add 2mol/L HCl, 2mol/L NaOH and saturated NH_4Cl solution into the other three test tubes, respectively. Observe whether the precipites dissolve or not.

(2) Compounds of Mn(Ⅳ)

① Preparation of MnO_2

Add 3 drops of 0.01mol/L $KMnO_4$ solution into a test tube, then add 5 drops of 0.1mol/L

MnSO$_4$ solution. Observe the formation of brown precipitate.

② Oxidizing activity of MnO$_2$

Add several drops of 6mol/L H$_2$SO$_4$ solution and a small amount of Na$_2$SO$_3$ crystal into a test tube with the MnO$_2$ prepared in test ①. Notice whether MnO$_2$ dissolves or does not.

(3) Compounds of Mn(Ⅵ)

① Preparation of K$_2$MnO$_4$

Add a small amount of MnO$_2$ and about an equal volume of KClO$_3$ solid into a dry test tube, then put two pellets of KOH into the tube. Mix evenly and heat the mixture to be melt. After cooling, add adequate amount of water to dissolve the products. A dark green K$_2$MnO$_4$ solution is obtained. Keep the K$_2$MnO$_4$ solution for the following experiments.

② Disproportionation of K$_2$MnO$_4$

Take 5 drops of K$_2$MnO$_4$ solution prepared in test ①. Add 5 drops of 2mol/L HAc solution, notice the color change of the solution and the formation of MnO$_2$. Write out the equation for this process.

③ Reducing activity of K$_2$MnO$_4$

Add several drops of aqueous chlorine into a test tube with 2 drops of K$_2$MnO$_4$ solution prepared in test ①. Notice the color of the solution. Write out the equation for this process.

(4) Compounds of Mn(Ⅶ)

Add 2 drops of 0.01mol/L KMnO$_4$ solution into each of the three test tubes, then add 2 drops of 2mol/L H$_2$SO$_4$ solution, 2 drops of water and 2 drops of 6mol/L NaOH solution into each test tube, respectively. Note what will occur after a small amount of Na$_2$SO$_3$ crystal is added into each tube. Write out the equations, and discuss the reduction products of KMnO$_4$ depending on different medium.

3. Iron compounds

(1) Preparation and properties of Fe(OH)$_2$

2ml of purified water acidified with 3 drops of 6mol/L H$_2$SO$_4$ solution is boiled in a test tube, then add some (NH$_4$)$_2$Fe(SO$_4$)$_2$ crystal. Shake to make the crystal dissolved. 3ml of 6mol/L boiled NaOH solution is added into the mixture above rapidly (Don't shake the tube). Note your observation. Let it rest for some time, then observe the color change of the precipitate.

Then prepare Fe(OH)$_2$ precipitate in the same way, and 2mol/L HCl solution is added into the precipitate rapidly. Observe whether the precipitate dissolves or not.

(2) Preparation and properties of Fe(OH)$_3$

Drop 2mol/L NaOH solution into a test tube with 1ml of 0.1mol/L FeCl$_3$ solution. Observe the color of the precipitate, then add 0.5ml of 2mol/L HCl solution into the tube. Observe whether the precipitate dissolves or not.

(3) Iron complexes

① Add 2 drops of 0.1mol/L FeCl$_3$ solution into a test tube, then add 2 drops of 0.1mol/L KI solution and one drop of starch solution. Add 2 drops of 0.1mol/L K$_3$[Fe(CN)$_6$] solution into another test tube with 2 drops of 0.1mol/L KI solution and one drop of starch solution. Note your observation of each test tube.

② Add 2 drops of 0.1mol/L $K_4[Fe(CN)_6]$ solution into a test tube, then add 2 drops of iodine solution and a small amount of $(NH_4)_2Fe(SO_4)_2$ crystal. Observe a blue precipitate or sol (vine's blue) (this reaction can be used to identify Fe^{2+}).

③ Add 2 drops of 0.1mol/L $FeCl_3$ solution into a test tube, then add one drop of 0.1mol/L $K_4[Fe(CN)_6]$ solution. Observe a dark blue precipitate or sol (prussian blue) (this reaction can be used to identify Fe^{3+}).

Notes

Chromium compounds are hazardous, especially Cr (Ⅵ) compounds, which are highly toxic. Cr (Ⅵ) compounds not only have a strong irritation to digestive tracts and skin but also have carcinogenic effects. Dosage should be minimized while operating. After the experiment the waste should be placed in a designated waste recycling containers.

Vocabulary

chromium hydroxide	$Cr(OH)_3$	chromium trichloride	$CrCl_3$
acetic acid	HAc	sulfuric acid	H_2SO_4
potassium chromate	K_2CrO_4	silver nitrate	$AgNO_3$
potassium dichromate	$K_2Cr_2O_7$	lead nitrate	$Pb(NO_3)_2$
barium chloride	$BaCl_2$	sodium sulfite	Na_2SO_3
manganese sulfate	$MnSO_4$	manganese dioxide	MnO_2
manganese hydroxide	$Mn(OH)_2$	sodium hydroxide	NaOH
potassium permanganate	K_2MnO_4	potassium permanganate	$KMnO_4$
ferric hydroxide	$Fe(OH)_3$	ferric chloride	$FeCl_3$
potassium iodide	KI	ferroushy droxide	$Fe(OH)_2$
potassium Ferricyanide	$K_3[Fe(CN)_6]$	ammonium chlorid	NH_4Cl
potassium ferrocyanide	$K_4[Fe(CN)_6]$	potassium chlorate	$KClO_3$

二、ds 区元素

【预习内容】

1. 在铜盐、银盐、锌盐和汞盐溶液中，分别加入 NaOH 溶液，是否均得到相应的氢氧化物？

2. 在铜盐、银盐、锌盐和汞盐溶液中，分别加入少量或过量氨水，产物分别是什么？

3. 在 $Hg_2(NO_3)_2$ 和 $Hg(NO_3)_2$ 溶液中，分别加入少量或过量的 KI 溶液，产物分别是什么？

4. 根据铜、汞的标准电极电势图，判断 Cu（Ⅰ）和 Hg（Ⅰ）的稳定性。

【思考题】

1. 为什么向 $CuSO_4$ 溶液中加入 KI 产生 CuI 沉淀，而向 $CuSO_4$ 溶液中加入 KCl 却得不到 CuCl 沉淀?

2. 如何使 Cu（Ⅰ）和 Cu（Ⅱ）稳定,它们相互转化的条件是什么?

【仪器与试剂】

试管,水浴。

固体:Hg_2Cl_2。

酸:H_2SO_4（2mol/L）。

碱:NaOH（2mol/L,6mol/L）,$NH_3 \cdot H_2O$（2mol/L）。

盐:$CuSO_4$（0.1mol/L）,$ZnSO_4$（0.1mol/L）,$AgNO_3$（0.1mol/L）,$Hg(NO_3)_2$（0.1mol/L）,KI（0.1mol/L,0.5mol/L）,$Na_2S_2O_3$（0.1mol/L）,KCl（0.1mol/L）,$HgCl_2$（0.1mol/L）,$Hg_2(NO_3)_2$（0.1mol/L）。

其他:甲醛溶液（5%）,葡萄糖溶液（10%）。

【操作步骤】

1. 铜、银、锌、汞氧化物或氢氧化物的生成和性质 在 4 支试管中分别加入 2 滴 0.1mol/L $CuSO_4$、$AgNO_3$、$ZnSO_4$、$Hg(NO_3)_2$ 溶液,然后依次加入 2mol/L NaOH,观察沉淀的颜色,并检验相应的氧化物或氢氧化物与酸、碱的作用。

2. 铜、银、锌、汞的配合物

(1) 氨配合物的生成 在 4 支试管中分别加入 2 滴 0.1mol/L $CuSO_4$、$AgNO_3$、$ZnSO_4$、$Hg(NO_3)_2$ 溶液,再分别加入 2mol/L $NH_3 \cdot H_2O$ 至沉淀生成,观察沉淀的颜色,继续加入过量的 2mol/L $NH_3 \cdot H_2O$,观察现象。

(2) 汞配合物的生成 在试管中加入 1 滴 0.1mol/L $Hg(NO_3)_2$ 溶液,再滴加少量 0.1mol/L KI 溶液,观察沉淀的颜色,继续加入过量的 0.5mol/L KI 溶液,至沉淀完全溶解。在所得的 $K_2[HgI_4]$ 溶液中滴加 6mol/L NaOH,使溶液呈碱性,即得到用来检验 NH_4^+（或 NH_3）的奈斯勒试剂,再加 1 滴 0.1mol/L NH_4Cl 溶液,可观察到红棕色沉淀,反应方程式如下:

$$NH_3 + 2[HgI_4]^{2-} + 3OH^- \rightleftharpoons \left[O \begin{array}{c} Hg \\ Hg \end{array} NH_2 \right] I \downarrow + 7I^- + 2H_2O$$

3. Cu（Ⅱ）的氧化性

(1) Cu_2O 的生成 在试管中加入 3 滴 0.1mol/L $CuSO_4$ 溶液,滴加过量的 6mol/L NaOH,使起初生成的蓝色沉淀溶解,然后加入 5 滴 5% 甲醛溶液,振荡后,水浴加热,观察沉淀的颜色变化。反应方程式如下:

$$2[Cu(OH)_4]^{2-} + HCHO = Cu_2O \downarrow + HCOOH + 4OH^- + 2H_2O$$

(2) CuI 的生成 在试管中加入 3 滴 0.1mol/L $CuSO_4$ 溶液,然后加入 5 滴 0.1mol/L KI 溶液,再滴加 0.1mol/L $Na_2S_2O_3$ 溶液至溶液颜色褪去（$Na_2S_2O_3$ 溶液不宜加得太多）,观察 CuI 沉淀的颜色,反应方程式如下:

$$2Cu^{2+} + 4I^- = 2CuI \downarrow + I_2$$

$$I_2 + 2S_2O_3^{2-} = 2I^- + S_4O_6^{2-}$$

在试管中加入 3 滴 0.1mol/L $CuSO_4$ 溶液,再加入 5 滴 0.1mol/L KCl 溶液,观察能否有 CuCl 沉淀生成?

4. Ag(Ⅰ)的氧化性(银镜反应) 在一支干净的试管中,加入 2ml 0.1mol/L $AgNO_3$ 溶液,滴加 2mol/L $NH_3·H_2O$ 至生成的沉淀刚好溶解,再往溶液中加几滴 10% 的葡萄糖溶液,摇匀后,将试管于水浴中加热,观察试管内壁上生成银镜。反应方程式如下:

$Ag(NH_3)_2^+ + C_5H_{11}O_5CHO + 2OH^- = Ag\downarrow + C_5H_{11}O_5COO^- + NH_4^+ + NH_3 + H_2O$

5. Hg(Ⅰ)、Hg(Ⅱ)的相互转化

(1) Hg(Ⅱ)的氧化性 同 Sn^{2+} 的鉴定反应。

(2) Hg(Ⅰ)的歧化反应

① 取 2 支试管,在一支试管中加入少量 Hg_2Cl_2 晶体,在另一支试管中加入 1 滴 0.1mol/L $HgCl_2$ 溶液,分别加入 2 滴 2mol/L $NH_3·H_2O$,观察沉淀的生成。

② 在试管中加 1 滴 0.1mol/L $Hg_2(NO_3)_2$ 溶液,再加入 1 滴 0.5mol/L KI 溶液,观察沉淀的颜色,继续加入过量的 KI 溶液,观察沉淀是否溶解?

【注意事项】

可溶性 Hg(Ⅱ)化合物有毒,使用时请注意。

Ⅱ. ds Block Elements

Preview

1. Can the corresponding hydroxides be obtained when NaOH solution is added to copper, silver, zinc and mercury salts solution, respectively?

2. What are the products when a little or excessive aqueous ammonia is added into copper, silver, zinc and mercury salts solution, respectively?

3. What are the products when a little or excessive KI solution is added into $Hg_2(NO_3)$ and $Hg(NO_3)_2$ solution, respectively?

4. Illustrate the stability of Cu(Ⅰ) and Hg(Ⅰ) using the diagrams of standard electrode potential of copper and mercury.

Questions

1. Why can CuI precipitate be generated when KI solution is added into $CuSO_4$ solution, while CuCl precipitate can't be generated when KI solution is replaced by KCl solution.

2. How to stabilize Cu(Ⅰ) or Cu(Ⅱ)? How can they interconvert?

Insruments and Chemicals

test tubes, water bath.

solid: Hg_2Cl_2.

acids: H_2SO_4 (2mol/L).

bases: NaOH (2mol/L, 6mol/L), $NH_3·H_2O$ (2mol/L).

salts: $CuSO_4$(0.1mol/L), $ZnSO_4$(0.1mol/L), $AgNO_3$(0.1mol/L), $Hg(NO_3)_2$(0.1mol/L), KI(0.1mol/L), $Na_2S_2O_3$(0.1mol/L), KCl(0.1mol/L), $HgCl_2$(0.1mol/L), $Hg_2(NO_3)_2$

(0.1mol/L).

others: formaldehyde solution (5%), glucose solution (10%).

Procedures

1. Formation and properties of oxides or hydroxides of copper, silver, zinc, mercury

Add 2 drops of 0.1mol/L $CuSO_4$, $AgNO_3$, $ZnSO_4$, $Hg(NO_3)_2$ solution into each of the four test tubes, then add 2mol/L NaOH solution into each tube. Observe the color of the precipitates and test whether the corresponding oxides or hydroxides can be dissolved in acid and alkali solution, respectively.

2. Complexes of copper, silver, zinc, mercury

(1) Formation of ammonia complexes

Add 2 drops of 0.1mol/L $CuSO_4$, $AgNO_3$, $ZnSO_4$, $Hg(NO_3)_2$ solution into each of the four test tubes, then drop 2mol/L aqueous ammonia into each tube, respectively, until the precipitates form. Observe the color of the precipitates. Then add excessive 2mol/L aqueous ammonia, note your observation.

(2) Preparation of mercury complexes

Add one drop of 0.1mol/L $Hg(NO_3)_2$ solution and a small amount of 0.1mol/L KI solution into a test tube. Observe the color of the precipitate. Then add excessive 0.5mol/L KI solution until the precipitate dissolves completely. Next add 6mol/L NaOH solution into the K_2HgI_4 solution until the mixture is alkaline. Nessler's reagent used to identify NH_4^+ (or NH_3) is obtained.

Combining the solution prepared above with one drop of 0.1mol/L NH_4Cl solution, red-brown precipitate is formed. The equation for this process is as follows.

$$NH_3 + 2[HgI_4]^{2-} + 3OH^- \rightleftharpoons [O\!\!\diagup_{Hg}^{Hg}\!\!\diagdown NH_2]I\downarrow + 7I^- + 2H_2O$$

3. Oxidizability of Cu(II)

(1) Formation and properties of Cu_2O

Add 3 drops of 0.1mol/L $CuSO_4$ solution into a test tube, then drop 6mol/L NaOH solution until the blue precipitate dissolves completely. Then 5 drops of 5% formaldehyde solution is added to the mixture. Shake and then heat in water bath. Observe the color change of the precipitate. The equation is as follows.

$$2[Cu(OH)_4]^{2-} + HCHO = Cu_2O\downarrow + HCOOH + 4OH^- + 2H_2O$$

(2) Formation of CuI

Add 3 drops of 0.1mol/L $CuSO_4$ solution and 5 drops of 0.1mol/L KI solution into a test tube, then add 0.1mol/L $Na_2S_2O_3$ solution into the test tube until the color of the solution just disappears (Control the amount of $Na_2S_2O_3$ solution). Observe the color of the CuI precipitate. The equations are as follows.

$$2Cu^{2+} + 4I^- = 2CuI\downarrow + I_2$$

$$I_2 + 2S_2O_3^{2-} = 2I^- + S_4O_6^{2-}$$

Add three drops of 0.1mol/L $CuSO_4$ solution into a test tube with 5 drops of 0.1mol/L KCl solution. Observe whether CuCl precipitate can generate.

4. Oxidizability of Ag(Ⅰ)

Add 2ml of 0.1mol/L $AgNO_3$ solution into a clean test tube, and drop 2mol/L aqueous ammonia into the tube until the precipitate just dissolves. Add 10% glucose solution into the mixture, stir thoroughly, and heat the tube with water bath. A shining mirror of silver is formed on the inner surface of the tube. The equation is as follows.

$$Ag(NH_3)_2^+ + C_5H_{11}O_5CHO + 2OH^- = Ag\downarrow + C_5H_{11}O_5COO^- + NH_4^+ + NH_3 + H_2O$$

5. Interconvertion of Hg(Ⅰ) and Hg(Ⅱ)

(1) Oxidizability of Hg(Ⅱ)

Refer to the identification of Sn^{2+}.

(2) Disproportionation of Hg(Ⅰ)

① Take two test tubes, and add a small amount of Hg_2Cl_2 crystal into one test tube, and add one drop of 0.1mol/L $HgCl_2$ solution into the other one, then drop 2 drops of 2mol/L aqueous ammonia into each tube. Observe the the formation of precipitates.

② Add one drop of 0.1mol/L $Hg_2(NO_3)_2$ solution and one drop of 0.5mol/L KI solution into a test tube. Observe the color of the precipitate. Then add excessive KI solution, observe whether the precipitate can dissolve.

Notes

Handle soluble Hg(Ⅱ) compounds carefully which are highly toxic.

Vocabulary

copper	铜	silver	银
zinc	锌	mercury	汞
silver nitrate	$AgNO_3$	zinc sulfate	$ZnSO_4$
mercuric nitrate	$Hg(NO_3)_2$	hydroxide	氢氧化物
complex	配合物	ammonia complexes	氨配合物
aqueous ammonia	氨水	ammonium ions	NH_4^+
Nessler's reagent	奈斯勒试剂	cuprous oxide	Cu_2O
cuprous Iodide	CuI	sodium thiosulfate	$Na_2S_2O_3$
formaldehyde	甲醛	glucose	葡萄糖
mercuric chloride	$HgCl_2$	mercurous chloride	Hg_2Cl_2
potassium tetraiodomercurate	K_2HgI_4		

（编写：赵兵，英文核审：夏丹丹）

实验十七 常见无机离子的鉴定反应

【预习内容】

1. 常见阳离子及阴离子的鉴定反应。
2. 定性分析的基本操作。

【仪器与试剂】

试管，烧杯（50ml），量筒（10ml），煤气灯，电炉，表面皿，点滴板，离心试管，离心机，镊子，玻璃棒，蓝色钴玻璃，绿色玻璃。

固体：$NaBiO_3$，MnO_2，铂丝，铜丝。

酸：HAc，HAc（2mol/L，6mol/L），盐酸，稀盐酸，HNO_3，稀HNO_3，H_2SO_4，稀H_2SO_4，对氨基苯磺酸溶液（0.02mol/L）。

碱：NaOH试液，NaOH（6mol/L），氨试液。

盐：$Hg_2(NO_3)_2$（试液），碱性碘化汞钾（$K_2HgI_4 + KOH$）（试液），四苯基硼酸钠溶液（$Na[B(C_6H_5)_4]$）（0.1%），NH_4Cl（试液），Na_2HPO_4（试液），$(NH_4)_2C_2O_4$（试液），$K_3[Fe(CN)_6]$（试液），$K_4[Fe(CN)_6]$（试液），NH_4SCN（试液），Na_2S（试液），KI（试液），K_2CrO_4溶液（5%），$BaCl_2$（试液），$Pb(Ac)_2$（试液），NH_4Ac（试液），$AgNO_3$（试液），$CaCl_2$（试液），$FeSO_4$（试液），$KMnO_4$（试液）。

其他：纯化水，氯（试液），淀粉（指示液），甲基红（指示液），茜素磺酸钠（指示液），邻二氮菲的乙醇溶液（1%），α-萘胺溶液（0.01mol/L），硫脲溶液（10%），红色石蕊试纸，碘化钾淀粉试纸，pH试纸，磷钼酸铵$\{(NH_4)_3[PMo_{12}O_{40}]\}$试纸，滤纸，碘液，三氯甲烷。

【操作步骤】

（一）常见阳离子的鉴定反应

1. 铵盐（NH_4^+）

（1）取供试品，加过量的NaOH试液后，加热，即分解，发生氨臭，遇用水湿润的红色石蕊试纸，能使之变蓝色，并能使$Hg_2(NO_3)_2$试液湿润的滤纸呈黑色。

$$NH_4^+ + OH^- = NH_3\uparrow + H_2O$$

$$4NH_3 + 2Hg_2(NO_3)_2 + H_2O = [O\underset{Hg}{\overset{Hg}{\diamond}}NH_2]NO_3 + 2Hg\downarrow + 3NH_4NO_3$$

（2）取供试品溶液，加碱性碘化汞钾（奈氏试剂）试液1滴，即生成红棕色沉淀。

$$NH_4^+ + 2[HgI_4]^{2-} + 4OH^- = [O\underset{Hg}{\overset{Hg}{\diamond}}NH_2]I\downarrow + 7I^- + 3H_2O$$

2. 钾盐（K^+）

（1）取铂丝，用盐酸湿润后，蘸取供试品，在无色火焰中燃烧，火焰即显紫色；但有少量的钠盐混存时，须隔蓝色玻璃透视，方能辨认。

（2）取供试品，加热炽灼除去可能杂有的铵盐，放冷后，加水溶解，再加0.1% $Na[B(C_6H_5)_4]$ 溶液与醋酸，即生成白色沉淀。

$$K^+ + [B(C_6H_5)_4]^- = K[B(C_6H_5)_4] \downarrow$$

3. 镁盐（Mg^{2+}）

（1）取供试品溶液，加氨试液，即生成白色沉淀；滴加 NH_4Cl 试液，沉淀溶解；再加 Na_2HPO_4 试液1滴，振摇，即生成白色沉淀。分离，沉淀在氨试液中不溶解。

$$Mg^{2+} + 2NH_3 \cdot H_2O = Mg(OH)_2 \downarrow + 2NH_4^+$$
$$Mg(OH)_2 + 2NH_4^+ = Mg^{2+} + 2NH_3 \uparrow + 2H_2O$$
$$Mg^{2+} + HPO_4^{2-} + NH_3 \cdot H_2O = MgNH_4PO_4 \downarrow + H_2O$$

（2）取供试品溶液，加 NaOH 试液，即生成白色沉淀。分离，沉淀分成2份，一份加过量的 NaOH 试液，沉淀不溶解；另一份中加碘试液，沉淀转成红棕色。

$$Mg^{2+} + 2OH^- = Mg(OH)_2 \downarrow$$

$Mg(OH)_2$ 可以强烈地吸附 I_2 而显红棕色。

4. 钙盐（Ca^{2+}）

（1）取铂丝，用盐酸湿润后，蘸取供试品，在无色火焰中燃烧，火焰即显砖红色。

（2）取供试品溶液（1→20），加甲基红指示液2滴，用氨试液中和，再滴加盐酸至恰呈酸性，加 $(NH_4)_2C_2O_4$ 试液，即生成白色沉淀；分离，沉淀不溶于醋酸，但可溶于稀盐酸。

$$Ca^{2+} + C_2O_4^{2-} = CaC_2O_4 \downarrow$$
$$CaC_2O_4 + 2H^+ = Ca^{2+} + H_2C_2O_4$$

5. 钡盐（Ba^{2+}）

（1）取铂丝，用盐酸湿润后，蘸取供试品，在无色火焰中燃烧，火焰即显黄绿色；通过绿色玻璃透视，火焰显蓝色。

（2）取供试品溶液，滴加稀硫酸，即生成白色沉淀；分离，沉淀在盐酸或硝酸中均不溶解。

$$Ba^{2+} + SO_4^{2-} = BaSO_4 \downarrow$$

6. 亚铁盐（Fe^{2+}）

（1）取供试品溶液，滴加 $K_3[Fe(CN)_6]$ 试液，生成深蓝色沉淀即滕氏蓝；分离，沉淀在稀盐酸中不溶，但加 NaOH 试液，即生成棕色沉淀。滕氏蓝被 NaOH 试液分解，并在空气中氧化，产生 $Fe(OH)_3$ 棕色沉淀。

$$Fe^{2+} + K^+ + [Fe(CN)_6]^{3-} = KFe[Fe(CN)_6] \downarrow （滕氏蓝）$$
$$KFe[Fe(CN)_6] + 3NaOH = Fe(OH)_2 + K^+ + OH^- + Na_3[Fe(CN)_6]$$
$$4Fe(OH)_2 + O_2 + 2H_2O = 4Fe(OH)_3 \downarrow$$

（2）取供试品溶液，加1%邻二氮菲的乙醇溶液数滴，即显深红色。

7. 铁盐（Fe^{3+}）

（1）取供试品溶液，滴加 $K_4[Fe(CN)_6]$ 试液，即生成深蓝色沉淀；分离，沉淀在稀盐酸中不溶，但加 NaOH 试液，即生成 $Fe(OH)_3$ 棕色沉淀。

$$Fe^{3+} + K^+ + [Fe(CN)_6]^{4-} = KFe[Fe(CN)_6] \downarrow \text{（普鲁士蓝）}$$

（2）取供试品溶液，滴加 NH_4SCN 试液，即显血红色。

$$Fe^{3+} + nSCN^- = [Fe(SCN)_n]^{3-n} \quad (n=1\sim6)$$

8. 亚锡盐（Sn^{2+}）

（1）取供试品的水溶液 1 滴，点于磷钼酸铵试纸上，试纸应显蓝色。

（2）酸性亚锡盐与磷钼酸铵反应，生成钼的低价氧化物，溶液为蓝色。

9. 锌盐（Zn^{2+}）

（1）取供试品溶液，加 $K_4[Fe(CN)_6]$ 试液，即生成白色沉淀；分离，沉淀在稀盐酸中不溶解。

$$2Zn^{2+} + K_4[Fe(CN)_6] = Zn_2[Fe(CN)_6] \downarrow + 4K^+$$

（2）取供试品制成中性或碱性溶液，加 Na_2S 试液，即生成白色沉淀。

$$Zn^{2+} + S^{2-} = ZnS \downarrow$$

10. 铝盐（Al^{3+}）

（1）取供试品溶液，滴加 NaOH 试液，即生成白色胶状沉淀；分离，沉淀能在过量的 NaOH 溶液中溶解。

$$Al^{3+} + 3OH^- = Al(OH)_3 \downarrow$$

$$Al(OH)_3 + OH^- = Al(OH)_4^-$$

（2）取供试品溶液，加氨试液至生成白色胶状沉淀，滴加茜素磺酸钠指示液数滴，沉淀即显樱红色。

11. 铋盐（Bi^{3+}）

（1）取供试品溶液，滴加 KI 试液，即生成红棕色溶液或暗棕色沉淀；分离，沉淀能在过量 KI 试液中溶解成黄棕色的溶液，再加水稀释，又生成橙色沉淀。

$$Bi^{3+} + 3KI = BiI_3 \downarrow + 3K^+$$

$$BiI_3 + KI\text{（过量）} = KBiI_4$$

$$KBiI_4 + H_2O = BiOI \downarrow + 2HI + KI$$

（2）取供试品溶液，用稀硫酸酸化，加 10% 硫脲溶液，即显深黄色。

$$Bi^{3+} + 3\underset{NH_2}{\underset{|}{C}}=S \rightleftharpoons \left[Bi\left(\underset{NH_2}{\underset{|}{C}}=S\right)_3\right]^{3+}$$

12. 铅盐（Pb^{2+}） 取供试品溶液，滴加 5% K_2CrO_4 溶液，即生成黄色沉淀。分离，沉淀分成 2 份，一份加 6mol/L 醋酸溶液，沉淀不溶解；另一份中加 6mol/L NaOH 溶液，沉淀溶解。

$$Pb^{2+} + CrO_4^{2-} = PbCrO_4 \downarrow$$

$$PbCrO_4 + 4OH^- = PbO_2^{2-} + 2H_2O + CrO_4^{2-}$$

13. 锰盐（Mn^{2+}） 取供试品溶液，滴加 6mol/L 硝酸使成酸性，加固体 $NaBiO_3$ 少许，溶液呈紫色。

$$2Mn^{2+} + 5NaBiO_3 + 14H^+ = 2MnO_4^- + 5Bi^{3+} + 5Na^+ + 7H_2O$$

（二）常见阴离子的鉴定反应

1. 硫酸盐（SO_4^{2-}）

（1）取供试品溶液，滴加 $BaCl_2$ 试液，即生成白色沉淀；分离，沉淀在盐酸或硝酸中均不溶解。

$$Ba^{2+} + SO_4^{2-} = BaSO_4 \downarrow$$

（2）取供试品溶液，滴加 Pb（Ac）$_2$ 试液，即生成白色沉淀；分离，沉淀在 NH_4Ac 试液或 NaOH 试液中可以溶解。

$$Pb^{2+} + SO_4^{2-} = PbSO_4 \downarrow$$

$$PbSO_4 + 2Ac^- = Pb(Ac)_2 + SO_4^{2-}$$

$$PbSO_4 + 4OH^- = PbO_2^{2-} + 2H_2O + SO_4^{2-}$$

（3）取供试品溶液，加盐酸，不生成白色沉淀（与硫代硫酸盐区别）。

2. 硫代硫酸盐（$S_2O_3^{2-}$） 取供试品溶液，滴加 $AgNO_3$ 试液，即生成白色沉淀，沉淀不稳定，立即发生水解反应，并且伴随明显的颜色变化，由白变黄、棕，最后变为黑色，生成 Ag_2S 沉淀。

$$2Ag^+ + S_2O_3^{2-} = Ag_2S_2O_3 \downarrow$$

$$Ag_2S_2O_3 + H_2O = Ag_2S \downarrow + 2H^+ + SO_4^{2-}$$

3. 草酸盐（$C_2O_4^{2-}$） 取供试品溶液，滴加 $CaCl_2$ 试液，即生成白色沉淀，加 6mol/L 醋酸，沉淀不溶解，加盐酸数滴，沉淀溶解。

$$Ca^{2+} + C_2O_4^{2-} = CaC_2O_4 \downarrow$$

$$CaC_2O_4 + 2H^+ = H_2C_2O_4 + Ca^{2+}$$

4. 氯化物（Cl^-）

（1）取供试品溶液，加稀硝酸酸化，滴加 $AgNO_3$ 试液，即生成白色凝乳状沉淀；分离，沉淀加氨试液即溶解，再加稀硝酸酸化后，沉淀复生成。如供试品为生物碱或其他有机碱的盐酸盐，须先加氨试液使成碱性，将析出的沉淀滤过除去，取滤液进行试验。

$$Ag^+ + Cl^- = AgCl\downarrow$$
$$AgCl + 2NH_3 = [Ag(NH_3)_2]^+ + Cl^-$$
$$[Ag(NH_3)_2]^+ + Cl^- + 2HNO_3 = AgCl\downarrow + 2NH_4NO_3$$

（2）取供试品少量，置试管中，加等量的 MnO_2，混匀，加硫酸湿润，缓缓加热，即发生氯气，能使用水湿润的碘化钾淀粉试纸显蓝色。

$$MnO_2 + 4H^+ + 2Cl^- = Mn^{2+} + Cl_2 + 2H_2O$$
$$Cl_2 + 2I^- = 2Cl^- + I_2$$

5. 溴化物（Br^-）

（1）取供试品溶液，滴加 $AgNO_3$ 试液，即生成淡黄色凝乳状沉淀；分离，沉淀能在氨试液中微溶，但在硝酸中几乎不溶。

$$Ag^+ + Br^- = AgBr\downarrow$$

（2）取供试品溶液，滴加氯试液，溴即游离，加三氯甲烷振荡，三氯甲烷层显黄色或红棕色。

$$Cl_2 + 2Br^- = 2Cl^- + Br_2$$

6. 碘化物（I^-）

（1）取供试品溶液，滴加 $AgNO_3$ 试液，即生成黄色凝乳状沉淀；分离，沉淀在硝酸或氨试液中均不溶解。

$$Ag^+ + I^- = AgI\downarrow$$

（2）取供试品溶液，加少量氯试液，碘即游离；如加三氯甲烷振荡，三氯甲烷层显紫色；如加淀粉指示液，溶液显蓝色。

$$Cl_2 + 2I^- = 2Cl^- + I_2$$

7. 亚硝酸盐（NO_2^-） 取供试品溶液，滴加 2mol/L 醋酸使成酸性，依次加入 0.02mol/L 对氨基苯磺酸溶液、0.01mol/L α-萘胺溶液，立即显红色。

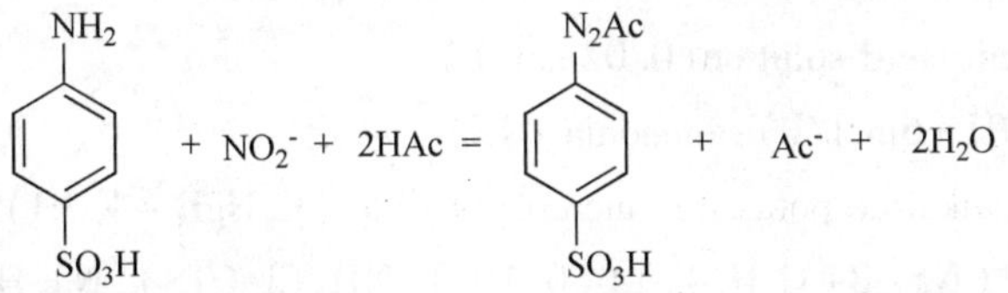

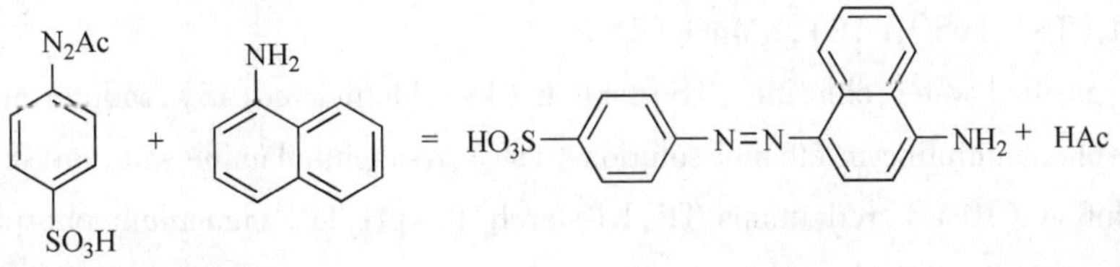

重氮化反应生成醋酸重氮盐，再与 α-萘胺发生偶联反应，产物是红色的偶氮染料。

8. 硝酸盐（NO_3^-）

（1）取供试品溶液，置试管中，加等量的硫酸，小心混合，冷却后，沿管壁加 $FeSO_4$ 试液，使成两液层，接界面显棕色。

$$NO_3^- + H_2SO_4 = HNO_3 + HSO_4^-$$
$$6FeSO_4 + 2HNO_3 + 3H_2SO_4 = 3Fe_2(SO_4)_3 + 2NO + 4H_2O$$
$$xFeSO_4 + yNO = xFeSO_4 \cdot yNO （棕色）$$

（2）取供试品溶液，加硫酸与铜丝（或铜屑），加热，即发生红棕色的蒸气。

$$NO_3^- + H_2SO_4 = HNO_3 + HSO_4^-$$
$$3Cu + 8HNO_3 = 3Cu(NO_3)_2 + 2NO + 4H_2O$$
$$2NO + O_2 = 2NO_2 \uparrow （红棕色）$$

（3）取供试品溶液，滴加 $KMnO_4$ 试液，紫色不应褪去（与亚硝酸盐区别）。

Experiment 17 Identification Reactions of the Common Inorganic Ions

Preview

1. Identification reactions of the common anions and cations.
2. Basic operations of qualitative analysis.

Instruments and Chemicals

test tube, beaker (50ml), cylinder(10ml), bunsen burner, electric stove, watch glass, spot plate, centrifuge tube, centrifuge machine, forceps, glass rod, blue cobalt glass, green glass.

solid: $NaBiO_3$, MnO_2, platinum wire, copper wire.

acid: HAc, HAc(2mol/L, 6mol/L), HCl, dilute HCl, HNO_3, dilute HNO_3, H_2SO_4, dilute H_2SO_4, p-aminobenzene sulfonic acid solution(0.02mol/L).

alkali: NaOH (TS), NaOH (6mol/L), ammonia (TS).

salts: $Hg_2(NO_3)_2$(TS), alkaline potassium mercuric iodide (K_2HgI_4 + KOH)(TS), sodium tetraphenylboron solution ($Na[B(C_6H_5)_4]$)(0.1%), NH_4Cl (TS), Na_2HPO_4 (TS), $(NH_4)_2C_2O_4$(TS), $K_3[Fe(CN)_6]$ (TS), $K_4[Fe(CN)_6]$ (TS), NH_4SCN (TS), Na_2S (TS), KI(TS), K_2CrO_4 solution (5%), $BaCl_2$ (TS), $Pb(Ac)_2$(TS), NH_4Ac (TS), $AgNO_3$ (TS), $CaCl_2$(TS), $FeSO_4$(TS), $KMnO_4$(TS).

others: purified water, chlorine (TS), starch (IS), Methyl red(IS), sodium alizarinsulfonate (IS), o-phenanthroline in ethanol solution (1%), α-naphthylamine solution(0.01mol/L), thiourea solution (10%), red litmus TP, KI-starch TP, pH TP, ammonium phosphomolybdate TP, filter paper, iodine solution, chloroform.

Procedures

Ⅰ. Identification reactions of the common cations

1. Ammonium Salts (NH_4^+)

(1) Heat a quantity of the substance being examined with an excess of NaOH TS; the characteristic odour of ammonia is perceived, the vapour turns moistened red litmus paper to blue and blackens a strip of filter paper moistened with $Hg_2(NO_3)_2$ TS.

$$NH_4^+ + OH^- = NH_3 \uparrow + H_2O$$

$$4NH_3 + 2Hg_2(NO_3)_2 + H_2O = [O{\underset{Hg}{\overset{Hg}{\diagup\hspace{-0.5em}\diagdown}}}NH_2]NO_3 + 2Hg \downarrow + 3NH_4NO_3$$

(2) To a solution of the substance being examined add 1 drop of alkaline mercuric potassium iodide (i.e. Nessler's reagent); a reddish brown precipitate is produced.

$$NH_4^+ + 2[HgI_4]^{2-} + 4OH^- = [O{\underset{Hg}{\overset{Hg}{\diagup\hspace{-0.5em}\diagdown}}}NH_2]I \downarrow + 7I^- + 3H_2O$$

2. Potassium Salts (K^+)

(1) Moisten the substance being examined with HCl on a platinum wire, it imparts a violet colour to a nonluminous flame. If sodium is also present, the yellow colour can be screened out by viewing through a cobalt glass plate.

(2) Ignite the substance being examined to remove any ammonium salt contaminated, cool, dissolve it in water, add HAc and a 0.1% solution of sodium tetraphenylborate; a white precipitate is produced.

$$K^+ + [B(C_6H_5)_4]^- = K[B(C_6H_5)_4] \downarrow$$

3. Magnesium Salts (Mg^{2+})

(1) Add ammonia TS to a solution of the substance being examined; a white precipitate is produced which redissolves on addition of NH_4Cl TS. Add a drop of Na_2HPO_4 TS and shake; a white precipitate insoluble in ammonia TS is produced.

$$Mg^{2+} + 2NH_3 \cdot H_2O = Mg(OH)_2 \downarrow + 2NH_4^+$$

$$Mg(OH)_2 + 2NH_4^+ = Mg^{2+} + 2NH_3 \uparrow + 2H_2O$$

$$Mg^{2+} + HPO_4^{2-} + NH_3 \cdot H_2O = MgNH_4PO_4 \downarrow + H_2O$$

(2) Add NaOH TS to a solution of the substance being examined; a white precipitate is produced which is insoluble in an excess of the reagent, filter, the precipitate is coloured reddish brown on addition of iodine TS.

$$Mg^{2+} + 2OH^- = Mg(OH)_2 \downarrow$$

$Mg(OH)_2$ can display reddish brown when iodine is adsorbed.

4. Calcium Salts (Ca^{2+})

(1) Moisten the substance being examined with HCl on a platinum wire, it imparts a yel-

lowish red colour to a nonluminous flame.

(2) Add 2 drops of methyl red IS to a solution of the substance being examined (1→20), neutralize with ammonia TS and then acidify with HCl. Add $(NH_4)_2C_2O_4$ TS; a white precipitate is produced which is soluble in HCl but insoluble in HAc.

$$Ca^{2+} + C_2O_4^{2-} = CaC_2O_4 \downarrow$$
$$CaC_2O_4 + 2H^+ = Ca^{2+} + H_2C_2O_4$$

5. Barium Salts (Ba^{2+})

(1) Moisten the substance being examined with HCl on a platinum wire, it imparts a yellowish green colour to a nonluminous flame, or a blue colour when viewed through a green glass plate.

(2) Add dilute H_2SO_4 to a solution of the substance being examined; a white precipitate is produced which is insoluble in HCl or HNO_3.

$$Ba^{2+} + SO_4^{2-} = BaSO_4 \downarrow$$

6. Ferrous Salts (Fe^{2+})

(1) Add $K_3[Fe(CN)_6]$ TS to a solution of substance being examined; a dark blue precipitate is formed which is insoluble in dilute HCl, it decomposes to form a brown precipitate on addition of NaOH TS.

$$Fe^{2+} + K^+ + [Fe(CN)_6]^{3-} = KFe[Fe(CN)_6] \downarrow \text{(Turnbull blue)}$$
$$KFe[Fe(CN)_6] + 3NaOH = Fe(OH)_2 + K^+ + OH^- + Na_3[Fe(CN)_6]$$
$$4Fe(OH)_2 + O_2 + 2H_2O = 4Fe(OH)_3 \downarrow$$

(2) To a solution of the substance being examined add a few drops of a 1% solution of o-phenanthroline in ethanol; a deep red colour is produced.

7. Ferric Salts (Fe^{3+})

(1) Add $K_4[Fe(CN)_6]$ TS to a solution of the substance being examined; a dark blue precipitate is formed which is insoluble in dilute HCl, it decomposes to form a brown precipitate on addition of NaOH TS.

$$Fe^{3+} + K^+ + [Fe(CN)_6]^{4-} = KFe[Fe(CN)_6] \downarrow \text{(Prussian blue)}$$

(2) Add NH_4SCN TS to a solution of substance being examined; a red colour is produced.

$$Fe^{3+} + nSCN^- = [Fe(SCN)_n]^{3-n} (n=1-6)$$

8. Stannous Salts (Sn^{2+})

One drop of an aqueous solution of the substance being examined turns $(NH_4)_3[PMo_{12}O_{40}]$ TP to blue.

The $(NH_4)_3[PMo_{12}O_{40}]$ can be reduced to molybdenum oxide at a lower price by acidic stannous salts, blue colour is produced.

9. Zinc Salts (Zn^{2+})

(1) Add $K_4[Fe(CN)_6]$ TS to a solution of the substance being examined; a white blue precipitate is formed which is insoluble in dilute HCl.

$$2Zn^{2+} + K_4[Fe(CN)_6] = Zn_2[Fe(CN)_6]\downarrow + 4K^+$$

(2) Add Na_2S TS to a neutral or alkaline solution of the substance being examined, a white precipitate is produced.

$$Zn^{2+} + S^{2-} = ZnS\downarrow$$

10. Aluminium Salts (Al^{3+})

(1) Add NaOH TS to a solution of the substance being examined; a gelatinous white precipitate appears which is soluble in an excess of NaOH TS.

$$Al^{3+} + 3OH^- = Al(OH)_3\downarrow$$
$$Al(OH)_3 + OH^- = Al(OH)_4^-$$

(2) Add ammonia TS to a solution of the substance being examined until a gelatinous white precipitate is formed, add a few drops of sodium alizarinsulfonate IS; the precipitate becomes cherry red in colour.

11. Bismuth Salts (Bi^{3+})

(1) Add KI TS to a solution of the substance being examined; a reddish brown colour or dark brown precipitate is produced, the precipitate is soluble in an excess of the reagent, forming a yellowish brown solution. Dilute the solution with water; an orange precipitate is produced.

$$Bi^{3+} + 3KI = BiI_3\downarrow + 3K^+$$
$$BiI_3 + KI(excess) = KBiI_4$$
$$KBiI_4 + H_2O = BiOI\downarrow + 2HI + KI$$

(2) Acidify a solution of the substance being examined with dilute H_2SO_4, drop a little 10% thiourea solution, an intense yellow colour is produced.

12. Lead Salts (Pb^{2+})

Add 5% solution of K_2CrO_4 to a solution of the substance being examined; a yellow precipitate is produced, which is soluble in NaOH solution (6mol/L) but insoluble in HAc(6mol/L).

$$Pb^{2+} + CrO_4^{2-} = PbCrO_4 \downarrow$$
$$PbCrO_4 + 4OH^- = PbO_2^{2-} + 2H_2O + CrO_4^{2-}$$

13. Manganese Salts (Mn^{2+})

Acidity a solution of the substance being examined with HNO_3 (6mol/L), add a small amount of solid $NaBiO_3$, solution appears dark purple colour.

$$2Mn^{2+} + 5NaBiO_3 + 14H^+ = 2MnO_4^- + 5Bi^{3+} + 5Na^+ + 7H_2O$$

II. Identification reactions of the common anions

1. Sulfates (SO_4^{2-})

(1) Add $BaCl_2$ TS to a solution of the substance being examined; a white precipitate is formed which is insoluble in HCl or HNO_3.

$$Ba^{2+} + SO_4^{2-} = BaSO_4 \downarrow$$

(2) Add $Pb(Ac)_2$ TS to a solution of the substance being examined; a white precipitate is formed which is soluble in NH_4Ac TS or NaOH TS.

$$Pb^{2+} + SO_4^{2-} = PbSO_4 \downarrow$$
$$PbSO_4 + 2Ac^- = Pb(Ac)_2 + SO_4^{2-}$$
$$PbSO_4 + 4OH^- = PbO_2^{2-} + 2H_2O + SO_4^{2-}$$

(3) Add HCl to a solution of the substance being examined; no white precipitate is produced (distinction from thiosulfates).

2. Thiosulfate ($S_2O_3^{2-}$)

Add $AgNO_3$ TS to solution of the substance being examined; a white precipitate is produced. Hydrolysis occurs instantly to the instable precipitate, forming Ag_2S. The color changes from white, yellow, brown, to black finally.

$$2Ag^+ + S_2O_3^{2-} = Ag_2S_2O_3 \downarrow$$
$$Ag_2S_2O_3 + H_2O = Ag_2S \downarrow + 2H^+ + SO_4^{2-}$$

3. Oxalates ($C_2O_4^{2-}$)

Add $CaCl_2$ TS to a solution of the substance being examined; a white precipitate is produced which is soluble in HCl and insoluble in 6mol/L HAc.

$$Ca^{2+} + C_2O_4^{2-} = CaC_2O_4 \downarrow$$
$$CaC_2O_4 + 2H^+ = H_2C_2O_4 + Ca^{2+}$$

4. Chlorides (Cl^-)

(1) Acidify a solution of the substance being examined with dilute HNO_3 and add $AgNO_3$ TS; a curdy, white precipitate is formed which is soluble in ammonia TS and reprecipitated on addition of dilute HNO_3. Alkaloids or organic bases should be removed by the addition of ammonia TS and filtration prior to the test.

$$Ag^+ + Cl^- = AgCl \downarrow$$
$$AgCl + 2NH_3 = [Ag(NH_3)_2]^+ + Cl^-$$
$$[Ag(NH_3)_2]^+ + Cl^- + 2HNO_3 = AgCl \downarrow + 2NH_4NO_3$$

(2) Mix a small quantity of the substance being examined with an equal part of MnO_2, moisten with H_2SO_4 and heat gently; chlorine is evolved which turns a strip of moistened KI-starch TP to blue.

$$MnO_2 + 4H^+ + 2Cl^- = Mn^{2+} + Cl_2 + 2H_2O$$

$$Cl_2 + 2I^- = 2Cl^- + I_2$$

5. Bromides (Br^-)

(1) Add $AgNO_3$ TS to a solution of the substance being examined; a curdy, pale yellow precipitate is formed which is slightly soluble in ammonia TS and practically insoluble in HNO_3.

$$Ag^+ + Br^- = AgBr \downarrow$$

(2) Add chlorine TS dropwise to a solution of the substance being examined; bromine is liberated, add chloroform and shake, a yellow to reddish brown colour is developed in the chloroform layer.

$$Cl_2 + 2Br^- = 2Cl^- + Br_2$$

6. Iodides (I^-)

(1) Add $AgNO_3$ TS to a solution of the substance being examined; a curdy, yellow precipitate is produced which is insoluble in HNO_3 or ammonia TS.

$$Ag^+ + I^- = AgI \downarrow$$

(2) Add chlorine TS dropwise to a solution of the substance being examined; iodine is liberated, add chloroform and shake, a violet colour is produced in the chloroform layer; if starch IS is added instead of chloroform, a blue colour is produced.

$$Cl_2 + 2I^- = 2Cl^- + I_2$$

7. Nitrites (NO_2^-)

Acidify a solution of the substance being examined with 2mol/L HAC. Add 0.02mol/L p-aminobenzene sulfonic acid solution, the add 0.01mol/L α-naphthylamine solution in turn, a red colour is immediately produced.

$$\text{H}_2\text{N-C}_6\text{H}_4\text{-SO}_3\text{H} + NO_2^- + 2HAc = \text{AcN}_2\text{-C}_6\text{H}_4\text{-SO}_3\text{H} + Ac^- + 2H_2O$$

$$\text{AcN}_2\text{-C}_6\text{H}_4\text{-SO}_3\text{H} + \text{naphthyl-NH}_2 = HO_3S\text{-C}_6H_4\text{-N=N-C}_{10}H_6\text{-NH}_2 + HAc$$

Diazonium acetate is produced by diazotization reaction. Coupling with α-naphthyl amine produces a red azo dye.

8. Nitrates (NO_3^-)

(1) Mix cautiously a solution of the substance being examined with an equal volume of H_2SO_4 and allow to cool, add $FeSO_4$ TS along the inner wall of the test tube; a brown ring is developed at the interface of the two layers.

$$NO_3^- + H_2SO_4 = HNO_3 + HSO_4^-$$
$$6FeSO_4 + 2HNO_3 + 3H_2SO_4 = 3Fe_2(SO_4)_3 + 2NO + 4H_2O$$
$$xFeSO_4 + yNO = xFeSO_4 \cdot yNO(brown)$$

(2) To a solution of the substance being examined add cautiously H_2SO_4 and metallic copper; a reddish brown fume is evolved on heating.

$$NO_3^- + H_2SO_4 = HNO_3 + HSO_4^-$$
$$3Cu + 8HNO_3 = 3Cu(NO_3)_2 + 2NO + 4H_2O$$
$$2NO + O_2 = 2NO_2 \uparrow (reddish\ brown)$$

(3) Add dropwise $KMnO_4$ TS to a solution of the substance being examined; the violet colour does not disappear (distinction from nitrites).

Vocabulary

cylinder	量筒	bunsen burner	煤气灯
electric stove	电炉	watch glass	表面皿
spot plate	点滴板	centrifuge tube	离心试管
centrifuge machine	离心机	forceps	镊子
glass rod	玻璃棒	blue cobalt glass	蓝色钴玻璃
green glass	绿玻璃	sodium bismuthate	铋酸钠
platinum wire	铂丝	manganese dioxide	二氧化锰
copper wire	铜丝	acetic acid	醋酸
hydrochloric acid	盐酸	dilute hydrochloric acid	稀盐酸
nitric acid	硝酸	sulfuric acid	硫酸
ammonia TS	氨试液	sodium sulfide	硫化钠
potassium iodide	碘化钾	barium chloride	氯化钡
lead acetate	醋酸铅	ammonium acetate	醋酸铵
silver nitrate	硝酸银	calcium chloride	氯化钙
ferrous sulfate	硫酸亚铁	chlorine TS	氯试液
starch IS	淀粉指示液	α – naphthylamine	α – 萘胺
thiourea	硫脲	pH TP	pH 试纸
red litmus TP	红色石蕊试纸		
ammonium phosphomolybdate	磷钼酸铵		
iodine	碘		
chloroform	三氯甲烷		
methyl red IS	甲基红指示液		
p-aminobenzene sulfonic acid	对氨基苯磺酸		

sodium hydroxide	氢氧化钠
mercurous nitrate	硝酸亚汞
alkaline potassium mercuric iodide TS 碱性碘化汞钾试液(K_2HgI_4 + KOH)	
sodium tetraphenylboron solution 四苯基硼酸钠溶液$\{Na[B(C_6H_5)_4]\}$	
ammonium chloride	氯化铵
disodium hydrogen phosphate	磷酸氢二钠
ammonium oxlate	草酸铵
potassium ferricyanide	铁氰化钾
potassium ferrocyanide	亚铁氰化钾
ammonium thiocyanate	硫氰酸氨
potassium chromate solution	铬酸钾溶液
potassium permanganate	高锰酸钾
sodium alizarinsulfonate	茜素磺酸钠
o-phenanthroline in ethanol solution	邻二氮菲的乙醇溶液
KI-starch TP	碘化钾淀粉试纸
test paper(TP)	试纸
test solution(TS)	试液
indicator solution(IS)	指示液

（编写：王鸿钢，英文核审：肖琰）

实验十八　样品分析

【预习内容】

1. 阴、阳离子初步试验的方法。
2. 相关阴、阳离子的鉴定方法。

【思考题】

1. 哪些阴离子会干扰SO_4^{2-}的检出？用什么方法可以消除干扰？
2. 在样品分析中为什么要进行初步试验？它是利用阴、阳离子哪些性质进行的？

【仪器与试剂】

离心机，点滴板。

酸：HCl（6mol/L，浓），H_2SO_4（3mol/L，6mol/L），HNO_3（6mol/L，浓）。

碱：NaOH（1mol/L，2mol/L，6mol/L），$NH_3 \cdot H_2O$（1mol/L，2mol/L，6mol/L）。

盐：$BaCl_2$（0.1mol/L），$AgNO_3$（0.1mol/L），$KMnO_4$（0.01mol/L），NH_4Cl（3mol/L）。

其他：I_2-淀粉混合液，KI-淀粉混合液，pH试纸，甲酸-甲酸铵缓冲溶液，醋酸-醋酸钠缓冲溶液，H_2S 气体。

【操作步骤】

样品由一种阳离子和一种阴离子组成。

1. 样品初步观察 观察样品的颜色、溶解性、结晶性状等。

2. 阴离子的分析

（1）阴离子试液的制备 取 0.5g 未知试样，加入 20ml 纯化水，溶解后所得溶液作分析用。

（2）阴离子初步试验

① 钡盐试验：取未知样试液 2 滴，加入 3 滴 0.1mol/L $BaCl_2$ 溶液，若有白色沉淀生成，表示可能有 SO_4^{2-}、$S_2O_3^{2-}$、PO_4^{3-}、SO_3^{2-}、CO_3^{2-} 等离子存在。然后离心分离，弃去上层清液，在沉淀中加入 6 滴 6mol/L HCl，若沉淀不能溶解，则表示有 SO_4^{2-} 存在。

② 银盐试验：取未知样试液 2 滴，加入 3 滴 0.1mol/L $AgNO_3$ 溶液，用 6 滴 6mol/L HNO_3 酸化后，再多加 1 滴～2 滴，充分搅拌。若生成黑色沉淀表示有 S^{2-} 或 $S_2O_3^{2-}$ 存在；若有黄色沉淀表示有 Br^- 或 I^- 存在，若有白色沉淀表示有 Cl^- 存在。

③ 还原性阴离子试验

（ⅰ）$KMnO_4$ 试验：取 4 滴未知样试液，用 4 滴 6mol/L H_2SO_4 酸化，然后加入 1 滴 0.01mol/L $KMnO_4$ 溶液，若紫色褪去，表示还原性离子如 S^{2-}、Br^-、I^-、SO_3^{2-}、NO_2^- 存在；若加热后紫色褪去，表示有 Cl^-、$C_2O_4^{2-}$ 离子存在。

（ⅱ）I_2 试验：取 2 滴未知样试液，用 2 滴 6mol/L H_2SO_4 酸化，然后加入 1 滴 I_2-淀粉混合液，若蓝色不消失，表示还原性阴离子不存在；若蓝色消失，可能存在 S^{2-}、$S_2O_3^{2-}$ 等离子。

④ 氧化性阴离子试验：取 2 滴未知样试液，用 2 滴 6mol/L H_2SO_4 酸化，然后加入 2 滴 KI-淀粉混合液，振荡试管，若溶液显蓝色，表示 NO_2^- 存在。

根据阴离子的初步试验，对照阴离子通性试验表，推断可能存在的阴离子。常见阴离子的检出见表 18-1。

表 18-1 常见阴离子的检出

通性试验 阴离子	钡盐试验		银盐试验		还原性试验		氧化性试验
	中性	加 HCl	中性	银盐沉淀加 HNO_3	$KMnO_4$ 试验	I_2 试验	KI 试验
SO_4^{2-}	白↓	↓不溶	—	—	—	—	—
SO_3^{2-}	白↓	↓溶解	白↓	↓溶解	紫红色褪去	蓝色褪去	—
CO_3^{2-}	白↓	↓溶解	白↓	↓溶解	—	—	—
PO_4^{3-}	白↓	↓溶解	黄↓	↓溶解	—	—	—
$C_2O_4^{2-}$	白↓	↓溶解	白↓	↓溶解	紫红色褪去△	—	—
S^{2-}	—	—	黑↓	↓溶解	紫红色褪去	蓝色褪去	—

续表

阴离子	通性试验	钡盐试验		银盐试验		还原性试验		氧化性试验
		中性	加 HCl	中性	银盐沉淀加 HNO_3	$KMnO_4$ 试验	I_2 试验	KI 试验
$S_2O_3^{2-}$		白↓	↓溶解 析出白色 S↓ + SO_2↑	白→黄 棕→黑	↓不溶	紫红色褪去	蓝色褪去	—
Cl^-		—	—	白↓	↓不溶	紫红色褪去△	—	—
Br^-		—	—	浅黄↓	↓不溶	紫红色褪去	—	—
I^-		—	—	黄↓	↓不溶	紫红色褪去	—	—
NO_2^-		—	—	白*↓	↓溶解	紫红色褪去	—	溶液变蓝
NO_3^-		—	—	—	—	—	—	—

—表示与试剂不发生反应 *表示离子浓度大时生成沉淀 △表示浓酸或加热时反应

（3）阴离子的确证试验 根据可能存在的阴离子进行确证试验，鉴定出存在的阴离子。

3. 阳离子的分析

（1）阳离子试液的制备 为防止阳离子的水解，阳离子应制成酸性溶液。

① 样品溶于水者：取样品 0.1g 于离心试管中，加水 5ml 溶解，如溶液呈碱性需加 6mol/L HNO_3 将其酸化。

② 样品溶于盐酸者：取样品 0.1g 于离心试管中，加浓 HCl 0.5ml 溶解，加水稀释到 5ml。

③ 样品溶于硝酸者：取样品 0.1g 于离心试管中，加浓 HNO_3 0.5ml 溶解，加水稀释到 5ml。

④ 样品溶于王水者：取样品 0.1g 于离心试管中，加王水 0.5ml 溶解，加水稀释到 5ml。

（2）阳离子的初步试验

①观察试液的颜色。

②pH 试验

（ⅰ）调节 pH 值在 1~3 之间：取阳离子试液 3 滴于离心试管中，如溶液为酸性，先用 6mol/L $NH_3·H_2O$ 中和到刚呈微碱性，加 3 滴 3mol/L NH_4Cl 溶液，再加 3 滴甲酸-甲酸铵缓冲溶液，使溶液的 pH 值在 1~3 之间，水浴加热并搅拌，观察有无沉淀生成，如有白色沉淀生成，则试样中可能存在 Bi^{3+}、Sn^{2+}、Sn^{4+}、Ag^+、Hg^{2+}、Pb^{2+} 离子（Pb^{2+} 离子浓度足够大时才生成沉淀）。

（ⅱ）调节 pH 值在 4~6 之间：取阳离子试液 3 滴于离心试管中，如溶液为酸性，先用 6mol/L $NH_3·H_2O$ 中和到刚呈微碱性，再加入 3 滴醋酸-醋酸钠缓冲溶液，使溶液 pH 值在 4~6 之间，水浴加热并搅拌，观察有无沉淀生成，如有沉淀生成则表示可能存在 Al^{3+}、Fe^{3+}、Bi^{3+}、Sb^{3+}、Sn^{2+}、Sn^{4+} 离子。

③氯化物试验：取阳离子试液 3 滴于离心试管中，加 1 滴 3mol/L HCl，如有沉淀

生成，则可能存在 Ag^+、Hg^{2+}、Pb^{2+} 离子（Pb^{2+} 离子浓度足够大时才生成沉淀）。

④硫酸盐试验：取阳离子试液 3 滴于离心试管中，加 2～4 滴 3mol/L H_2SO_4，并水浴加热，观察有无沉淀生成，如有沉淀生成，则可能存在 Pb^{2+}、Sr^{2+}、Hg^{2+}（Ca^{2+} 离子浓度足够大时，开始生成沉淀）。

⑤氢氧化物试验

（ⅰ）等量氢氧化钠处理：取阳离子试液 3 滴于离心试管中，加 1mol/L NaOH 使溶液刚呈碱性，如有沉淀生成，观察生成沉淀的颜色，根据阳离子初步试验表判断可能存在的阳离子。

（ⅱ）过量氢氧化钠处理：于上述所得沉淀中，加入过量 NaOH 溶液，如沉淀消失，则可能存在 Al^{3+}、Pb^{2+}、Zn^{2+}、Cr^{3+}、Sb^{3+}、Sn^{2+}、Sn^{4+} 等两性离子。

⑥氨水试验

（ⅰ）等量氨水处理：取阳离子试液 3 滴于离心试管中，加 1mol/L $NH_3 \cdot H_2O$ 使溶液呈碱性，如有沉淀生成，观察生成沉淀的颜色，根据阳离子初步试验表判断可能存在的阳离子。

（ⅱ）过量氨水处理：在（ⅰ）中所得沉淀中，加入过量 $NH_3 \cdot H_2O$，如沉淀消失，表示生成了配离子，根据溶液的颜色，参照阳离子初步试验表判断可能存在的阳离子。

⑦硫化物试验

（ⅰ）用 HCl 调节试液酸度 0.3mol/L，将离心试管置沸水水浴中加热 1～2 分钟，通 H_2S 气体，观察有无沉淀生成。

如有沉淀生成，则继续通入 H_2S 直至沉淀完全，离心分离。根据沉淀颜色判断可能存在的离子［沉淀留作下述（ⅲ）硝酸处理］。

如无沉淀生成，此时溶液可直接进行下述（ⅱ）的试验。

（ⅱ）在碱性溶液中生成硫化物沉淀：取溶液（ⅰ）于离心试管中，加入 6mol/L $NH_3 \cdot H_2O$ 至呈碱性（必要时再通入 H_2S），观察有无沉淀生成，如生成沉淀，根据沉淀颜色判断可能存在的离子。

（ⅲ）硝酸处理沉淀

于沉淀（ⅰ）中加入 6mol/L HNO_3 及 0.5mol/L KNO_2 溶液各 2 滴，置沸水浴中加热并搅拌，如果沉淀转变为白色，表示可能有 Sb^{3+}、Sn^{2+}、Sn^{4+} 离子存在。如沉淀为黑色、不溶于 HNO_3，表示有 Hg^{2+} 离子存在。

根据阳离子的初步试验，对照阳离子初步试验表，推测可能存在的阳离子。

（3）阳离子的确证试验　根据初步试验结果，对可能存在的阳离子进行确证试验。

4. 总结推断，做出结论。

【书写报告】

样品分析实验报告

分析人_____

样品号		样品外观	
液性（pH）		样品名称	

阴离子初步试验

阴离子	BaCl₂		AgNO₃		还原性试验		氧化性试验
	中性	加 HCl	中性	加 HNO₃	KMnO₄	I₂试验	KI 试验
初步结论							

阴离子确证鉴定反应

阴离子	试剂	现象	鉴定反应方程式
结论			

阳离子分析报告自拟。

Experiment 18　Analysis of Samples

Preview

1. Anionic and cationic preliminary test methods.
2. Determination of some common anions and cations.

Questions

1. Which kind of anions can interfere with the detection of SO_4^{2-}? How to eliminate the interference?

2. What is the purpose of preliminary tests in determination of sample? What properties of anions and cations are used in preliminary tests?

Instruments and Chemicals

centrifugal machine, spot plate.

acids: HCl (3mol/L, 6mol/L, concentrated), H_2SO_4(3mol/L, 6mol/L), HNO_3(6mol/L, concentrated).

bases: NaOH (1mol/L, 2mol/L, 6mol/L), $NH_3 \cdot H_2O$(1mol/L, 2mol/L, 6mol/L).

salts: $BaCl_2$(0.1mol/L), $AgNO_3$(0.1mol/L), $KMnO_4$(0.01mol/L), NH_4Cl(3mol/L).

others: I_2-amylum mixed solution, KI-amylum mixed solution, pH test paper, formic acid-ammonium formate buffer solution, H_2S gas.

Procedures

The sample consists of one type cation and one type anion.

1. Preliminary observation of samples

Note the color, the solubility and crystallization properties of the sample.

2. Determination of anion

(1) Preparation of solution

Take 0.5g of unknown sample dissolved in 20ml of purified water. Keep the solution for anion analysis.

(2) Preliminary tests for anions

① Barium salts test

Take 2 drops of the sample from procedure (1), and add 3 drops of 0.1mol/L $BaCl_2$ solution. If white precipitate appears, it indicates the presence of SO_4^{2-}, $S_2O_3^{2-}$ or CO_3^{2-}. Centrifuge and discard the supernate, then add 6 drops 6mol/L HCl solution to the precipitate. If the precipitate is insoluble, it indicates the presence of SO_4^{2-}.

② Silver salts test

Take 2 drops of the sample from procedure (1), add 3 drops of 0.1mol/L $AgNO_3$ solution. After acidified with 6 drops 6mol/L HNO_3 solution (excessive 1 drop or 2 drops HNO_3 solution is added), the mixture is stirred well. The anion can be pre-estimated according to the color of the precipitate. If black precipitate appears, it indicates the presence of S^{2-} or $S_2O_3^{2-}$. If yellow precipitate appears, it indicates the presence of Br^- or I^-. The formation of white precipitates indicates the presence of Cl^-.

③ Reducing anion test

(ⅰ) $KMnO_4$ test

Take 4 drops of the sample solution acidified with 4 drops of 6mol/L H_2SO_4 solution, then add 1 drop of 0.01mol/L $KMnO_4$ solution. If the red violet color of the solution disappears, it indicates the presence of reducing anions such as S^{2-}, Br^-, I^-, SO_3^{2-}, NO_2^-. If the red violet color disappears after heating, it indicates the presence of Cl^- or $C_2O_4^{2-}$ (heated if necessary).

(ⅱ) I_2 test

Take 2 drops of the sample solution acidified with 2 drops of 6mol/L H_2SO_4 solution, then add 1 drop of I_2-amylum mixed solution. If the blue color of the solution does not disappear, it proves that the reductive anion doesn't not exist. If the blue color disappears, it indicates the presence of S^{2-} or $S_2O_3^{2-}$.

④ Oxidizing anions test

Take 2 drops of the sample solution acidified with 2 drops of 6mol/L H_2SO_4 solution, then add 2 drops of KI-amylum mixed solution, and shake well. If the blue color appears, it indicates the presence of NO_2^-.

According to preliminary tests for anions and common properties test table of anions below, the anion can be pre-estimated. Detection of common anions is listed in Table 18-1.

Table 18-1   Detection of common anions

test \ anion	Barium salt test		Silver salts test		Reducing test		oxidizing test
	neutral	add HCl	neutral	add HNO$_3$	KMnO$_4$ test	I$_2$ test	KI test
SO_4^{2-}	white ↓	↓ insoluble	—	—	—	—	—
SO_3^{2-}	white ↓	↓ soluble	white ↓	↓ soluble	red violet color disappears	blue color disappears	—
CO_3^{2-}	white ↓	↓ soluble	white ↓	↓ soluble	—	—	—
PO_4^{3-}	white ↓	↓ soluble	yellow ↓	↓ soluble	—	—	—
$C_2O_4^{2-}$	white ↓	↓ soluble	white ↓	↓ soluble	red violet color disappears △	—	—
S^{2-}	—	—	black ↓	↓ soluble	red violet color disappears	blue color disappears	—
$S_2O_3^{2-}$	white ↓	↓ soluble S↓ + SO$_2$↑	white→yellow brown→black	↓ insoluble	red violet color disappears	blue color disappears	—
Cl^-	—	—	white ↓	↓ insoluble	red violet color disappears △	—	—
Br^-	—	—	light yellow ↓	↓ insoluble	red violet color disappears	—	—
I^-	—	—	yellow ↓	↓ insoluble	red violet color disappears	—	—
NO_2^-	—	—	white * ↓	↓ soluble	red violet color disappears	—	blue color appears
NO_3^-	—	—	—	—	—	—	—

— no reaction with the reagents

* the precipitate is formed while a large ion concentration

△ the reaction occurs while the presence of concentrated acid or heating

(3) Determinate tests for anions

Carry out determinate tests for possible anions and then identify the anion.

3. Determination of cations

(1) Preparation of solution with unknown cations

Cations solution should be acidified in order to prevent the hydrolysis of cations.

① Samples dissolved in water: dissolve 0.1g of samples in 5ml of water in a centrifugal test tube. If the solution is alkaline, acidify it with 6mol/L HNO$_3$.

② Samples dissolved in hydrochloric acid: dissolve 0.1g of samples in 0.5ml concentrated HCl in a centrifugal test tube, then dilute the solution by adding purified water to 5ml.

③ Samples dissolved in nitric acid: dissolve 0.1g of samples in 0.5ml concentrated HNO$_3$

in a centrifugal test tube, then dilute the solution by adding purified water to 5ml.

④ Samples dissolved in aqua regia: dissolve 0.1g of samples in 0.5ml aqua regia in a centrifugal test tube, then dilute the solution by adding purified water to 5ml.

(2) Preliminary tests for the cations

① The color of the sample solution.

② pH test

(ⅰ) Adjust pH in the range of 1 to 3

Take 3 drops of sample solution into a centrifugal test tube. If the solution is acidic, neutralize it with 6mol/L $NH_3 \cdot H_2O$ until the solution is just weakly alkaline. Add 3 drops of 3mol/L NH_4Cl solution, then add 3 drops of formic acid-ammonium formate buffer solution until the pH value of the solution is in the range of 1 to 3, heat in water bath, stir, and observe whether the precipitate appears. The formation of white precipitates indicates the possible presence of Bi^{3+}, Sn^{2+}, Sn^{4+}, Ag^+, Hg^{2+} or Pb^{2+} in sample solution (a precipitate appears when Pb^{2+} concentration is large enough).

(ⅱ) Adjust pH in the range of 4 to 6

Take 3 drops of sample solution into a centrifugal test tube. If the solution is acidic, neutralize it with 6mol/L $NH_3 \cdot H_2O$ until the solution is just weakly alkaline. Add 3 drops of 3mol/L NH_4Cl solution, then add 3 drops of formic acid-ammonium formate buffer solution until the pH value of the solution is in the range of 4 to 6. Heat in water bath, stir, and observe whether the precipitate appears. The formation of precipitates indicates the possible presence of Al^{3+}, Fe^{3+}, Bi^{3+}, Sb^{3+}, Sn^{2+}, Sn^{4+} in sample solution.

③ Chloride test

Take 3 drops of sample into a centrifugal test tube, and add one drop of 3mol/L HCl solution. The formation of the precipitates indicates the presence of Ag^+, Hg^{2+}, Pb^{2+} (a precipitate appears when Pb^{2+} concentration is high enough).

④ Sulphate Test

Take 3 drops of sample solution into a centrifugal test tube, add 2 to 4 drops of 3mol/L H_2SO_4 solution. Heat in water bath, and observe whether the precipitate appears. The formation of the precipitates indicates the presence of Pb^{2+}, Sr^{2+}, Hg^{2+} (a precipitate appears when Ca^{2+} concentration is high enough).

⑤ Hydroxide test

(ⅰ) Stoichiometric sodium hydroxide

Take 3 drops of sample solution into a centrifugal test tube, add 1mol/L NaOH until the solution is just alkaline. If precipitates appear, note the color of precipitates. Determine the presence of possible cations according to preliminary tests table of cations.

(ⅱ) Excessive sodium hydroxide

Add excessive sodium hydroxide solution to the precipitate obtained in procedure (ⅰ). If the precipitate is dissolved, it indicates the presence of amphoteric ions such as Al^{3+}, Pb^{2+},

Zn^{2+}, Cr^{3+}, Sb^{3+}, Sn^{2+}, Sn^{4+}.

⑥ Ammonia test

(ⅰ) Stoichiometric Ammonia

Take 3 drops of sample solution to a centrifugal test tube, add 1mol/L $NH_3 \cdot H_2O$ until the sample solution is just alkaline. If the precipitate appears, note the color of the precipitate. Determine the presence of possible cations according to preliminary tests table of cations.

(ⅱ) Excessive Ammonia

Add excessive $NH_3 \cdot H_2O$ solution into the precipitates obtained in step (ⅰ). If the precipitates disappear, it indicates the formation of complex ions. Determine the presence of possible cations according to the color of the solution and preliminary tests table of cations.

⑦ Sulfide Test

(ⅰ) Add hydrochloric acid into the sample solution until the concentration of hydrogen ion is 0.3mol/L. After the centrifuge tube heated in a boiling water bath for 1min to 2min, pass into H_2S gas, and observe whether the precipitate appears.

If the precipitate appears, continue to pass into H_2S gas until precipitate completely, and then centrifuge. Determine the possible ion according to the color of the precipitates (the precipitates are kept for treatment with nitric acid in step (ⅲ)).

If no precipitates, operate test in step (ⅱ).

(ⅱ) The formation of sulfide precipitates in alkaline solution

Add 6mol/L $NH_3 \cdot H_2O$ into a centrifugal test tube with the solution stocked from step (ⅰ) until the solution is alkaline (pass into H_2S gas if necessary), and observe whether the precipitate appears. Determine the possible ion according to the color of the precipitates.

(ⅲ) Handling precipitate with nitric acid

Add 2 drops of 6mol/L HNO_3 and 2 drops of 0.5mol/L KNO_2 solution into the precipitate stocked from step (ⅰ), heat in boiling water bath and stir. If the precipitate turns white, it indicates the presence of Sb^{3+}, Sn^{2+}, Sn^{4+}. The formation of black HgS, which is insoluble in HNO_3 solution, indicates the presence of Hg^{2+}.

According to preliminary tests of cations, check common properties test table of cations, possible cations can be determined.

(3) Confirmatory test of cation

Carry out confirmatory test for possible cations and then identify the relevant cantions.

4. Summary and conclusions.

Data record and processing

Lab report for sample analysis

Analysts _____

sample number		samples appearance	
pH		sample name	

Preliminary tests for the anions

anion	BaCl$_2$ test		AgNO$_3$ test		Reducing test		Oxidating test
	neutral	add HCl	neutral	add HNO$_3$	KMnO$_4$ test	I$_2$ test	KI test
preliminary conclusions							

Identification of anions

anion	reagents	phenomenon	equation for identification
conclusions			

Design the report of analysis of cation by yourself.

（编写：赵兵，英文核审：夏丹丹）

第五章　无机化合物制备实验

Chapter 5　Preparation Experiments of Inorganic Compounds

实验十九　药用氯化钠的制备及杂质限度检查

【预习内容】
1. 掌握药用氯化钠的制备原理和方法。
2. 练习和巩固称重、减压过滤、蒸发浓缩等基本操作。
3. 初步了解药品的鉴别、检查方法。

【思考题】
1. 加 $BaCl_2$ 之前为何要将粗 NaCl 的饱和溶液，加热至近沸腾？
2. 如何检查沉淀是否完全？
3. 重结晶的目的是什么？
4. 加 HCl 溶液的目的是什么？
5. 蒸发浓缩的过程中为什么要不断搅拌？
6. 炒盐的目的是什么？

【实验原理】
1. 药用氯化钠的制备　食盐是溶于水的固态物质、对于其中所含杂质的消除方法基本是：

（1）机械杂质、如泥沙等不溶性杂质，可以通过将 NaCl 溶于水后用过滤方法除去。

（2）一些能溶解的杂质可根据其性质借助于化学方法除去，如加入 $BaCl_2$ 溶液可使 SO_4^{2-} 生成 $BaSO_4$ 沉淀，加 Na_2CO_3 溶液可使 Ca^{2+}、Mg^{2+}、Fe^{3+}、Ba^{2+} 等离子生成难溶物沉淀，然后滤去。

（3）少量可溶性杂质如 Br^-、I^-、K^+ 等离子；可根据溶解度的不同，在重结晶时，使其残留在母液中而除去。

2. 鉴别试验　是被检药品组成离子的特征试验（这是指的是 NaCl 的组成离子 Na^+、Cl^-）。

3. 杂质限度试验　钡盐、硫酸盐、钾盐、钙镁盐的限度检查，是根据沉淀反应原理，样品管和标准管在相同条件下进行比浊试验，样品管不得比标准管更深。

4. 重金属　系指 Pb、Bi、Cu、Hg、Sb、Sn、Co、Zn 等金属的离子，它们在一定

条件下能与 H_2S 或 Na_2S 作用而显色。中国药典规定是在弱酸性条件下进行，用稀醋酸调节。实验证明，在 pH＝3 时，PbS 沉淀最完全。反应式为：
$$Pb^{2+} + H_2S \rightarrow PbS\downarrow + 2H^+$$
重金属检查是在相同条件下进行比色试验。

【仪器与试剂】

电子天平，电热套，真空泵，烧杯（100ml、250ml、50ml），布氏漏斗，抽滤瓶，蒸发皿，铂丝，比色管。

固体：NaCl（粗盐），可溶性淀粉，MnO_2，$(NH_4)_2S_2O_8$。

酸：HCl（2mol/L，0.02mol/L），HNO_3，H_2SO_4（稀），HAc（稀）。

碱：NaOH（2mol/L，0.02mol/L），Na_2CO_3（饱和），$NH_3 \cdot H_2O$（TS），硫代乙酰胺（TS）。

盐：$BaCl_2$（25%），K_2CO_3（15%），焦锑酸钾（TS），$AgNO_3$（TS），$Na_2S_2O_3$（0.1mol/L），$(NH_4)_2C_2O_4$（TS），NH_4SCN（30%），氯胺T溶液（0.01%），四苯硼钠溶液，醋酸盐缓冲液。

其他：KI-淀粉试纸，溴麝香草酚蓝指示液，苯酚红混合液，淀粉混合液，太坦黄溶液（0.05%）。

【操作步骤】

1. 粗氯化钠提纯

（1）粗氯化钠溶解　称取50g粗盐于250ml烧杯中，加入约200ml自来水，加热搅拌使其溶解。

（2）除 SO_4^{2-}　加热溶液至近沸腾，边搅拌边滴加25% $BaCl_2$ 试液约6~8ml，以除去溶液中 SO_4^{2-}。取清液检验 SO_4^{2-} 除尽后，继续加热煮沸数分钟，抽滤除去沉淀。

（3）除 Ca^{2+}、Mg^{2+}、Fe^{3+} 和过量的 Ba^{2+}　将滤液转移至另外一个250ml干净烧杯中，加热至沸腾。边搅拌边滴加饱和 Na_2CO_3 溶液，并用2mol/L NaOH 溶液，调溶液 pH 值10~11左右，以除去溶液中 Ca^{2+}、Mg^{2+}、Fe^{3+} 和过量的 Ba^{2+}。取清液检验 Ba^{2+} 等除尽后，继续加热煮沸数分钟，抽滤除去沉淀。

（4）除去剩余的 CO_3^{2-} 和 OH^-　将滤液移入蒸发皿内，在搅拌下滴加2mol/L HCl 调节溶液的 pH 值3~4。

（5）蒸发浓缩　加热蒸发浓缩上述溶液，并不断搅拌至糊状稠液为止，趁热抽滤至干，弃去滤液（含 K^+、Br^-、I^- 等离子），得到固体 NaCl。

（6）重结晶　将滤得的 NaCl 固体移入干净蒸发皿，加适量水不断搅拌至完全溶解为止，如上法进行蒸发浓缩，纯 NaCl 结晶析出，K^+、Br^-、I^- 等离子留在溶液中，趁热抽滤，抽干。

将固体 NaCl 置干燥蒸发皿中，小火炒干，冷却至室温，称重，计算产率。

2. 鉴别反应

（1）钠盐

① 取铂丝，用 HCl 湿润后，蘸取 NaCl 溶液，在无色火焰中燃烧，火焰即显鲜

黄色。

② 取 NaCl 100mg，置 10ml 试管中，加水 2ml 溶解，加 15% K_2CO_3 溶液，加热至沸，应不得有沉淀；加焦锑酸钾试液 4ml，加热至沸；置冰水中冷却，必要时，用玻棒摩擦试管壁，应有致密沉淀生成。

（2）氯化物

① 取 NaCl 溶液，加 HNO_3 使成酸性后，加 $AgNO_3$ 试液，即生成白色凝乳状沉淀；分离，沉淀加氨试液即溶解，再加稀 HNO_3 酸化后，沉淀复生成。

② 取 NaCl 少量，置试管中，加等量的 MnO_2，混匀，加 H_2SO_4 湿润，缓缓加热，即发生氯气，能使湿润的 KI–淀粉试纸显蓝色。

3. 检查 成品 NaCl 需进行以下各项质量检查试验。

（1）溶液的澄清度　取本品 5.0g，加水至 25ml 溶解后，溶液应澄清。

（2）酸碱度　取本品 5.0g，加水 50ml 溶解后，加溴麝香草酚蓝指示液 2 滴，如显黄色，加 NaOH 滴定液（0.02mol/L）0.10ml，应变为蓝色；如显蓝色或绿色，加 HCl 滴定液（0.02mol/L）0.20ml，应变为黄色。

NaCl 为强酸强碱所生成的盐：在水溶液中应呈中性。但在制备过程中，可能夹杂少量酸或碱，所以药典把它限制在很小范围。溴麝香草酚蓝指示液的变色范围是 pH 6.6~7.6，由黄色到蓝色。

（3）碘化物　取本品 5.0g，置瓷蒸发皿内，滴加新配制的淀粉混合液适量使晶粒湿润，置日光下（或日光灯下）观察，5 分钟内晶粒不得显蓝色痕迹。

（4）溴化物　取本品 2.0g，置 100ml 量瓶中，加水溶解并稀释至刻度，摇匀，精密量取 5ml，置 10ml 比色管中，加苯酚红混合液（pH 4.7）2.0ml 和 0.01% 的氯胺 T 溶液（临用新制）1.0ml，立即混匀，准确放置 2 分钟，加 0.1mol/L 的 $Na_2S_2O_3$ 溶液 0.15ml，用水稀释至刻度，摇匀，作为供试品溶液；另取标准 KBr 溶液 5.0ml，置 10ml 比色管中，同法制备，作为对照溶液。取对照溶液和供试品溶液，照紫外–可见分光光度法，以水为空白，在 590nm 处测定吸光度，供试品溶液的吸光度不得大于对照溶液的吸光度（0.01%）。

（5）硫酸盐　取本品 5.0g，加水溶解使成约 40ml（溶液如显碱性可滴加 HCl 使成中性）；溶液如不澄清，应滤过；置 50ml 纳氏比色管中，加稀 HCl 2ml，摇匀，即得供试溶液。取标准 K_2SO_4 溶液 1.0ml，置 50ml 纳氏比色管中，加水使成约 40ml，加稀 HCl 2ml，摇匀，即得对照溶液。两管内分别加入 25% $BaCl_2$ 溶液 5ml，用水稀释至 50ml，充分摇匀，放置 10 分钟，同置黑色背景上，从比色管上方向下观察，比较。供试品溶液若出现沉淀，与对照品溶液相比不得更浓（0.002%）。

（6）钡盐　取本品 4.0g，加水 20ml 溶解后，滤过，滤液分为两等份，一份中加稀 H_2SO_4 2ml，另一份中加水 2ml，静置 15 分钟，两液应同样澄清。

（7）钙盐　取本品 2.0g，加水 10ml 使溶解，加氨试液 1ml，摇匀，加 $(NH_4)_2C_2O_4$ 试液 1ml，5 分钟内不得发生浑浊。

（8）镁盐　取本品 1.0g，加水 20ml 使溶解，加 NaOH 试液 2.5ml 与 0.05% 太坦黄溶液 0.5ml，摇匀，生成的颜色与标准镁溶液 1.0ml 用同一方法制成的对照液比较，不得更深（0.001%）。

（9）钾盐　取本品 5.0g，加水 20ml 溶解后，加稀 CH_3COOH 2 滴，加四苯硼钠溶液 2ml，加水使成 50ml，如显浑浊，与标准 K_2SO_4 溶液 12.3ml 用同一方法制成的对照液比较，不得更浓（0.02%）。

（10）铁盐　取本品 5.0g，加水溶解使成 25ml，移置 50ml 纳氏比色管中，加稀 HCl 4ml 与 $(NH_4)_2S_2O_8$ 50mg，用水稀释使成 35ml，加 30% NH_4SCN 溶液 3.0ml，再加水至 50ml，摇匀；如显色，立即与标准铁溶液 1.5ml 用一方法制成的对照液比较，不得更深（0.0003%）。

（11）重金属　含重金属不得过 2ppm。

取 25ml 纳氏比色管三支。甲管中加标准铅溶液 1ml 与醋酸盐缓冲液（pH 3.5）2ml 后，加水稀释成 25ml。乙管中加入 5g 本品，加水 20ml 溶解，加醋酸盐缓冲液（pH 3.5）2ml 与水适量使成 25ml。丙管中加入 5g 本品，加水 20ml 溶解，加标准铅溶液 1ml 与醋酸盐缓冲液（pH 3.5）2ml 后，加水稀释成 25ml。在甲、乙、丙三管中分别加硫代乙酰胺试液各 2ml，摇匀，放置 2 分钟，同置白纸上，自上向下透视，当丙管中显出的颜色不浅于甲管时，乙管中显出的颜色与甲管比较，不得更深。

附注

（1）淀粉混合液的配制　取可溶性淀粉 0.25g，加水 2ml，搅匀，再加沸水至 25ml，随加随搅拌，放冷，加 0.025mol/L H_2SO_4 溶液 2ml、$NaNO_2$ 试液 3 滴与水 25ml，混匀。

（2）苯酚红混合液的配制　取 $(NH_4)_2SO_4$ 25mg，加水 230ml，加 2mol/L NaOH 溶液 105ml，加 2mol/L CH_3COOH 溶液 135ml，摇匀，加苯酚红溶液（取苯酚红 33mg，加 2mol/L NaOH 溶液 1.5ml，加水溶解并稀释至 100ml，摇匀，即得）25ml，摇匀，调节 pH 至 4.7。

（3）标准 KBr 溶液的配制　精密称取在 105℃ 干燥至恒重的 KBr 30mg，加水使溶解成 100ml，摇匀，精密量取 1ml，置 100ml 量瓶中，用水稀释至刻度，摇匀，即得（每 1ml 溶液相当于 2μg 的 Br）。

（4）标准 K_2SO_4 溶液的配制　精密称取 K_2SO_4 0.181g，置 1000ml 量瓶中，加水适量使溶解并稀释至刻度，摇匀，即得（每 1ml 相当于 100μg 的 SO_4^{2-}，相当于 81μg 的 K^+）。

（5）标准镁溶液的配制　精密称取在 800℃ 炽灼至恒重的 MgO 16.58mg，加 HCl 2.5ml 与水适量使溶解成 1000ml，摇匀，即得（每 1ml 相当于 10μg 的 Mg）。

（6）四苯硼钠溶液的配制　取四苯硼钠 1.5g，置乳钵中，加水 10ml 研磨后，再加水 40ml，研匀，用质密的滤纸滤过，即得。

（7）标准铁溶液的制备　精密称取硫酸铁铵 $[FeNH_4(SO_4)_2 \cdot 12H_2O]$ 0.863g，置 1000ml 量瓶中，加水溶解后，加 H_2SO_4 2.5ml，用水稀释至刻度，摇匀，作为贮备液。临用前，精密量取贮备液 10ml，置 100ml 量瓶中，加水稀释至刻度，摇匀，即得（每 1ml 相当于 10μg 的 Fe）。

（8）标准铅溶液的制备　称取 $Pb(NO_3)_2$ 0.1599g，置 1000ml 量瓶中，加 HNO_3

5ml 与水 50ml 溶解后，用水稀释至刻度，摇匀，作为贮备液。精密量取贮备液 10ml，置 100ml 量瓶中，加水稀释至刻度，摇匀，即得（每 1ml 相当于 10μg 的 Pb）。本液仅供当日使用。

配制与贮存用的玻璃容器均不得含铅。

Experiment 19　Preparation of Medicinal Sodium Chloride and Examination of the Limits of Impurities

Preview

1. To master the principles and method to prepare medicinal sodium chloride.

2. To practice and consolidate the basic operations of weighing, vacuum filtration, evaporation and condensation.

3. To learn the methods of identification and test of sodium chloride.

Questions

1. Why is the saturated solution of crude sodium chloride heated to boil before the addition of $BaCl_2$ solution?

2. How to verify that any impurity was completely precipitated?

3. What is the purpose of re-crystallization processing?

4. What is the purpose of the addition of HCl solution?

5. Why continuously stirring was required during the evaporation and condensation operation?

6. Why the product should be fried?

Principles

1. Preparation of medicinal sodium chloride

Sodium chloride is soluble in aqueous solution, the impurities in sodium chloride can be removed with the following procedures:

(1) The water-insoluble impurities can be removed by filtration.

(2) Some soluble impurities can be removed by precipitation basing on their chemical properties. For example, sulfate can be separated as $BaSO_4$ by the addition of $BaCl_2$ solution; Ca^{2+}, Mg^{2+}, Ba^{2+}, Fe^{3+}, etc. can be separated as insoluble precipitates by addition of Na_2CO_3 solution.

(3) Some low content soluble impurities, such as Br^-, I^-, K^+, can be removed by re-crystallization. After re-crystallization, these impurities retained in the mother liquid and can be separated thereof.

2. Identification of sodium chloride

The identification of sodium chloride include the specific tests on the composition ions, ie.

Na^+ and Cl^-.

3. Limit tests of impurities

The limit tests of barium, sulfate, potassium, calcium and magnesium are carried out using comparison tubes by addition of corresponding precipitate agents. Under the same conditions, any opalescence produced in the test solutions should not be pronounced than that of the reference solution.

4. Heavy metals

Heavy metals, including the ions of Pb, Bi, Cu, Hg, Sb, Sn, Co, Zn, and other metals, can be colored by sulfide ion under the specified test conditions. The test of heavy metallic impurities is carried out by comparing the color of solutions with the corresponding reference solutions under the same conditions.

Instruments and Chemicals

electronic balance, electric jacket, vacuum pump, beakers (100ml, 250ml, 50ml), Buchner funnel, filter flask, evaporating dish, platinum wire, Nessler cyltnder.

solid: NaCl (crude), MnO_2, soluble starch, $(NH_4)_2S_2O_8$.

acid: HCl (2mol/L, 0.02mol/L), HNO_3, H_2SO_4(dilute), HAc(dilute).

alkali: NaOH (2mol/L, 0.02mol/L), Na_2CO_3 (saturated), $NH_3 \cdot H_2O$ (TS), thioacetamide (TS).

salt: $BaCl_2$ (25%), K_2CO_3 (15%), potassium pyroantimonate (TS), $AgNO_3$ (TS), $Na_2S_2O_3$ (0.1mol/L), $(NH_4)_2C_2O_4$ (TS), NH_4SCN (30%), chloramine test solution (0.01%), sodium tetraphenylboron solution, acetate(BS).

others: starch-iodide paper, bromothymol blue (TS), phenol red TS, starch mixture, titan yellow solution (0.05%).

Procedures

1. Purification of the crude salt

(1) Dissolving of the crude salt

Weigh 50g of crude salt in a 250ml beaker, add about 200ml tap water, heat and stir to dissolve.

(2) Removing of SO_4^{2-}

Heat the above solution to boil, add about 6—8ml 25% $BaCl_2$ solution with stirring. Take sample to test whether SO_4^{2-} is removed completely. Heat until it boils for several minutes, remove any insoluble substances by suction filtration.

(3) Removing of Ca^{2+}, Mg^{2+}, Fe^{3+} and excess Ba^{2+}

Transfer the filtrate to another 250ml beaker, heat to boil. Add saturated Na_2CO_3 solution with stirring, adjust pH to 10—11 with 2mol/L NaOH. Take sample to test whether Ba^{2+} is removed completely. Heat until it boils for several minutes, remove any insoluble substances by suction filtration.

(4) Removing residuary CO_3^{2-} and OH^-

Transfer the filtrate to an evaporating dish, adjust pH to 3—4 with 2mol/L HCl.

(5) Evaporation and Condensation

Heat the above solution with continuously stirring to evaporate the excess of water, filter the concentrated solution while it is still warm, cool to room temperature and obtain the solid NaCl.

(6) Re-crystallization

Place the solid NaCl to an evaporating dish, add purified water, heat and stir to make it dissolve, evaporate and condense to precipitate the crystals of pure NaCl, while remain the ions in solution, such as K^+, Br^-, I^-.

Transfer the pure NaCl to an evaporating dish, fried with mild fire. Cool to room temperature, weigh and calculate the yield (%).

2. Identification

(1) Sodium Salts

① Moisten the substance being examined with hydrochloric acid on a platinum wire; it imparts an intense yellow colour to a nonluminous flame.

② Prepare a solution to contain 100mg of the sodium chloride in 2ml of water. Add 2ml of 15% potassium carbonate, and heat to boiling. No precipitate is formed. Add 4ml of potassium pyroantimonate TS, and heat to boiling. Allow to cool in ice water and, if necessary, rub the inside of the test tube with a glass rod. A dense precipitate is formed.

(2) Chlorides

① Acidify a solution of the sodium chloride with nitric acid and add silver nitrate TS; a curdy, white precipitate is formed which is soluble in ammonia TS and reprecipitated on addition of nitric acid.

② Mix a small quantity of the sodium chloride with an equal part of manganese dioxide, moisten with sulfuric acid and heat gently; chlorine is evolved which turns a strip of moistened starch-iodide paper to blue.

3. Test

The product should be checked with the limit tests described as below.

(1) Clarity of solution

Dissolve 5.0g of NaCl in carbon dioxide-free water and dilute with the same solvent to 25ml. This solution is clear and colorless.

(2) Acidity or alkalinity

Dissolve 5.0g of NaCl in carbon dioxide-free water and dilute with the same solvent to 50ml. Add 2 drops of bromothymol blue TS; not more than 0.2ml of 0.02mo/L hydrochloric acid or 0.1ml of 0.02mol/L sodium hydroxide is required to change the color of this solution.

The content of free acid or alkali, if there is any, in medicinal NaCl should not be more than the limit.

Sodium chloride is a salt formed by the strong acid and strong alkali, the solution of sodium chloride in water should be neutral. However, a small amount of acid or base may be mixed in the preparation process, the Pharmacopoeia specified a very narrow pH range. And the pH range of bromothymol blue is 6.6—7.6, yellow to blue.

（3）Iodides

Moisten 5.0g of NaI in a porcelain evaporating dish by the dropwise addition of a freshly prepared starch mixture. Examine the mixture in daylight; no particle shows any trace of blue colour within 5 minutes.

（4）Bromides

Dissolve 2.0g of NaBr in water and dilute with water to 100ml, mix well. Accurately transfer 5ml of the solution to a 10ml Nessler cylinder, add 2.0ml of phenol red TS pH 4.7, and 1.0ml of 0.01% chloramine test solution (prepare before use), mix immediately. After 2 minutes, add 0.15ml of 0.1mol/L sodium thiosulfate, mix, dilute with water to 10.0ml, and mix. The absorbance of this solution measured at 590nm, using water as the reference, is not greater than that of a standard solution, concomitantly prepared, using 5.0ml of a solution containing 3.0mg of potassium bromide per L and proceeding as above, starting with the addition of 2.0ml of phenol red test solution pH 4.7 (0.01%).

（5）Sulfate

Reference preparation — Transfer a 1ml of potassium sulfate standard solution to a 50ml nessler cylinder, dilute with water to about 40ml, add 2ml of dilute hydrochloric and mix well.

Test preparation — Dissolve 5g of NaCl in about 40ml of water, neutralize the solution with hydrochloric acid and filter if necessary. Transfer the solution to a 50ml Nessler cylinder, add 2ml of diluted hydrochloric acid and mix well.

Procedure — To each of the Nessler cylinders described above add 5ml 25% $BaCl_2$ solution, dilute with water to 50ml and mix thoroughly, allow to stand for 10 minutes and compare the opalescence produced by viewing down the vertical axis of the cylinder against a black background. Any opalescence produced is no more pronounced than that of the reference solution (0.002%).

（6）Barium

Dissolve 4.0g $BaCl_2$ in 20ml of water, filer and divide the filtrate into two equal portions. To one portion add 2ml of diluted sulfuric acid and to the other add 2ml of water. The solutions are equally clear after standing for 15 minutes.

（7）Calcium

Dissolve 2.0g $CaCl_2$ in 10ml water, add 1ml of ammonia TS, mix well, add 1ml of ammonium oxalate TS. No opalescence is produced within 5minutes.

（8）Magnesium

Dissolve 1.0g $MgCl_2$ in 20ml of water, add 2.5ml of sodium hydroxide TS and 0.5ml of 0.05% titan yellow solution and mix well. Any color produced is not more intense than that of a reference solution using 1.0ml of standard magnesium solution (0.001%).

(9) Potassium

Dissolve 5.0g KCl in 20ml of water. Transfer the solution to a 50ml Nessler cylinder, add 2 drops of dilute acetic acid and 2ml of sodium tetraphenylboron solution, dilute with water to 50ml and mix thoroughly. Any opalescence produced is not more pronounced than that of a reference solution using 12.3ml of standard potassium sulfate solution (0.02%).

(10) Iron

Dissolve 5.0g $FeCl_2$ in 25ml of water. Transfer the solution to a 50ml Nessler cylinder, add 4ml of dilute hydrochloric acid and 50mg of ammononium persulfate. Dilute with water to about 35ml, add 3ml of 30% ammonium thiocyanate solution and sufficient water to produce 50ml, mix well. Any color produced is not more intense than that of a reference solution using 1.5ml of iron standard solution (0.0003%).

(11) Heavy Metals

Carry out the limit test as bellow: not more than 0.002%.

Refenence preparation— Transfer 1ml of lead standard solution to a 25ml nessler cylinder, add 2ml of acetate BS (pH 3.5), dilute with water to 25ml.

Test preparation — Dissolve 5.0g this product in 20ml of water. Transfer the solution to a 25ml Nessler cylinder, add 2ml of acetate BS (pH 3.5), dilute with water to 25ml.

Monitor preparation — Dissolve 5.0g this product in 20ml of water. Transfer the solution to a 25ml Nessler cylinder, add 1ml of lead standard solution and 2ml of acetate BS (pH 3.5), dilute with water to 25ml.

Procedure — To each of the three cylinder containing the Refenence Preparation, the Test Preparation, and the Monitor Preparation, add 2ml of thioacetamide TS, mix well, allow to stand for 2 minutes, and view downward over a white surface: the color of the solution from the Test Preparation is not darker than that of the solution from the Refenence Preparation, and the intensity of the color of the Monitor Preparations equal to or greater than that of the Refenence Preparation.

Vocabulary

Medicinal	药用的	sodium chloride	氯化钠
iodides	碘化物	vacuum filtration	抽滤
bromides	溴化物	filtrate	滤液(*n.*), 过滤(*v.*)
sulfate	硫酸盐	filter	滤器(*n.*), 过滤(*v.*)
calcium	钙盐	magnesium	镁盐
barium	钡盐	evaporation	蒸发
potassium	钾盐	evaporating dish	蒸发皿
iron	铁盐	condensation	浓缩
heavy metals	重金属离子	crystallization	结晶
dissolve	溶解	re-crystallization	重结晶

dilute	稀(释)的	precipitate	沉淀
pronounced	显著的	clarity	澄清度
opalescence	乳(白)光	Nessler cylinder	纳氏比色管
nonluminous flame	无色火焰	TS	试液(缩略语)
potassium pyroantimonate	焦锑酸钾		
BS	缓冲溶液(缩略语)		
platinum wire	铂丝	Standard solution	标准溶液
potassium carbonate	碳酸钾	starch	淀粉
bromothymol blue	溴麝香草酚蓝	phenol red	苯酚红
titan yellow	太坦黄	sodium thiosulfate	硫代硫酸钠溶液
potassium bromide	溴化钾	magnesium oxide	氧化镁
lead nitrate	硝酸铅	ammononium persulfate	过硫酸铵
ammonium thiocyanate	硫氰酸铵	ferric ammonium sulfate	硫酸亚铁铵
thioacetamide	硫代乙酰胺	sodium tetraphenylboron	四苯硼钠溶液
manganese dioxide	二氧化锰		

(编写：王绍宁，英文核审：肖琰)

实验二十　硫酸亚铁铵的制备

【预习内容】

1. 了解铁（Ⅱ）化合物的性质。
2. 查阅硫酸亚铁和硫酸铵在不同温度下的溶解度。
3. 了解产品检验的分析方法。

【思考题】

1. 在制备硫酸亚铁铵过程中，溶液为什么必须呈酸性？
2. 反应为什么在70℃进行？
3. 为什么用乙醇洗涤晶体？
4. 检验产品等级时，为什么要用不含氧纯化水？
5. 硫酸亚铁铵的收率怎么计算？计算收率时以铁的量还是以硫酸铵的量为准？

【实验原理】

铁屑易溶于稀 H_2SO_4 中，生成 $FeSO_4$：即：

$$Fe + H_2SO_4 = FeSO_4 + H_2\uparrow$$

硫酸亚铁铵 $[(NH_4)_2SO_4 \cdot FeSO_4 \cdot 6H_2O]$ 是用等摩尔量的 $FeSO_4$ 和 $(NH_4)_2SO_4$ 在水溶液中反应制得的，其反应方程式为：

$$FeSO_4 + (NH_4)_2SO_4 + 6H_2O = (NH_4)_2SO_4 \cdot FeSO_4 \cdot 6H_2O$$

硫酸亚铁铵在水中溶液度较小，它比较稳定，不易被空气氧化，所以是化学分析中常用的基准物之一。

【仪器与试剂】

电子天平，烧杯（50ml），锥形瓶（100ml），布氏漏斗，抽滤瓶（250ml），比色管（25ml），蒸发皿，恒温水浴锅，三颈瓶（250ml），加料漏斗。

HCl（3mol/L），H_2SO_4（0.2mol/L、3mol/L），KSCN（25%），铁屑，$(NH_4)_2SO_4$（s），Na_2CO_3（10%）。

【操作步骤】

1. 制备

（1）铁屑表面油污的去除 称取铁屑5.6g放在小烧杯中，加入10% Na_2CO_3的溶液40ml，小火加热10分钟，倾出碱液，用水洗净铁屑，备用。

（2）$FeSO_4$的制备 往盛有铁屑的锥形瓶内加入40ml的3mol/L的H_2SO_4，在水浴上加热，使铁屑与H_2SO_4完全反应，应不时地往锥形瓶中加水及H_2SO_4溶液，以补充被蒸发掉的水分，趁热减压过滤，并用少量热水洗涤滤渣，保留滤液。

（3）$(NH_4)_2SO_4 \cdot FeSO_4 \cdot 6H_2O$的制备 将14.0g $(NH_4)_2SO_4$置于50ml烧杯中，加入20ml的0.2mol/L的H_2SO_4，在70℃恒温水浴中使之溶解，然后将溶液转移到加料漏斗中待用。将（2）中制备的$FeSO_4$滤液转移至250ml三颈瓶中并加热至70℃，开始滴加$(NH_4)_2SO_4$溶液，边加边搅拌，滴完后反应5分钟。反应完毕后将反应液转移至蒸发皿中，蒸发浓缩至有晶膜出现，冷却至室温析出结晶，减压抽滤，用95%乙醇5ml洗涤一次，将所得结晶称重，计算收率。

2. 产品检验

（1）铁（Ⅲ）的限量分析 称量1g产品置于50ml烧杯中，用15ml不含氧的纯化水溶解之，然后转移到25ml比色管中，加入2ml的3mol/L HCl和1ml的25% KSCN溶液，继续加不含氧的纯化水至25ml刻度，摇匀，与标准溶液进行目视比色，确定产品的等级。

（2）Fe^{3+}标准溶液的制备 分别取5ml、10ml、20ml 0.01mg/ml的Fe^{3+}溶液，处理方法与产品相同。

Ⅰ级试剂：0.05mg（每1g产品中Fe^{3+}的含量小于0.05mg）

Ⅱ级试剂：0.10mg（每1g产品中Fe^{3+}的含量介于0.05~0.10mg）

Ⅲ级试剂：0.20mg（每1g产品中Fe^{3+}的含量介于0.10~0.20mg）

3. 含量分析 称取1.5g样品，精确至0.0001g，溶于100ml无氧的纯化水中，加5ml硫酸、2ml磷酸及200ml无氧的纯化水，摇匀，用高锰酸钾标准滴定液（c = 0.02mol/L）滴定至溶液呈粉红色，保持30秒。

六水合硫酸亚铁铵的质量分数ω，数值以"%"表示，按下式计算：

$$\omega = c \cdot V \cdot M / 1000m \times 100$$

式中，c为$KMnO_4$标准滴定液的浓度（mol/L），V为$KMnO_4$标准滴定液体积（ml），M为$(NH_4)_2SO_4 \cdot FeSO_4 \cdot 6H_2O$的摩尔质量（g/mol），$m$为样品质量（g）。

附注

表 20-1　不同温度时硫酸铵溶解度

温度（℃）	溶解度（g/100g 水）	温度（℃）	溶解度（g/100g 水）
1	70.6	50	—
10	73	60	88
20	75.4	70	89.6
30	78	80	95.3
40	81	100	103.3

表 20-2　不同温度时硫酸亚铁的溶解度

温度（℃）	溶解度（g/100g 水）	温度（℃）	溶解度（g/100g 水）
1	15.6	57	—
10	20.5	65	—
20	26.5	70	50.9
30	32.9	80	43.6
40	40.2	90	37.3
50	48.6		

表 20-3　不同温度时硫酸亚铁铵溶解度

温度（℃）	溶解度（g/100g 水）	温度（℃）	溶解度（g/100g 水）
1	12.5	40	33
10	—	50	40
20	—	60	—
30	—		

Experiment 20　Preparation of Ferrous Ammonium Sulphate

Preview

1. To understand the properties of ferrous compound.
2. To look up the solubility values of $FeSO_4$ and $(NH_4)_2SO_4$ at different temperatures.
3. To understand the analytical method of impurity limitation.

Questions

1. Why should the solution be acidity during preparation of $(NH_4)_2SO_4 \cdot FeSO_4 \cdot 6H_2O$?
2. Why should the reaction temperature be kept at 70℃?
3. Why is the ethanol used to wash crystal after filtration?
4. Why should the oxygen-free purified water be used when check out the purity grade of the product?
5. How to calculate the yield of $(NH_4)_2SO_4 \cdot FeSO_4 \cdot 6H_2O$? Which is the basis of sub-

stance to this calculation, Fe or $(NH_4)_2SO_4$?

Principles

Iron is easily soluble in dilute H_2SO_4, and generates $FeSO_4$.

$$Fe + H_2SO_4 = FeSO_4 + H_2 \uparrow$$

$FeSO_4$ reacts with $(NH_4)_2SO_4$ in equimolar ratio in aqueous solution to get $FeSO_4 \cdot (NH_4)_2SO_4 \cdot 6H_2O$. The reaction equation is written as:

$$FeSO_4 + (NH_4)_2SO_4 + 6H_2O = (NH_4)_2SO_4 \cdot FeSO_4 \cdot 6H_2O$$

$FeSO_4 \cdot (NH_4)_2SO_4 \cdot 6H_2O$ is less soluble in water, stable and not easy to be oxidized in air. So it is one of the commonly used standard substances in chemical analysis.

Instruments and Chemicals

electronic balance, beaker (50ml), erlenmeyer flask (100ml), Buchner funnel, suction flask (250ml), colorimetric tube (25ml), evaporating dish, thermostatic water bath, three-necked flask (250ml), charging hopper.

HCl (3mol/L), H_2SO_4 (0.2mol/L, 3mol/L), KSCN (25%), iron powder, $(NH_4)_2SO_4$ (s), Na_2CO_3 (10%).

Procedures

1. Preparation

(1) Wash the grease stain on iron powder

5.6g of iron powder and 40ml of 10% Na_2CO_3 are put into a beaker. Then heat with small fire for 10 minutes. Remove the alkaline solution by tilt-pour process and wash the iron powder clean with purified water to prepare for use.

(2) Preparation of $FeSO_4$

Add 40ml of 3mol/L H_2SO_4 into the erlenmeyer flask with the treated iron powder, and then heat it in the water bath at 70℃ until the reaction is completed. At the meantime, add water and H_2SO_4 solution to supplement the evaporated water. Filter under vaccum while it is hot and wash the residue with small amount of hot water. Reserve the filtrate.

(3) Preparation of $FeSO_4 \cdot (NH_4)_2SO_4 \cdot 6H_2O$

14.0g of $(NH_4)_2SO_4$ and 20ml of 0.2mol/L H_2SO_4 are put into a 50ml beaker, which is heated at 70℃ in the water bath until the $(NH_4)_2SO_4$ is dissolved. And then transfer the solution into the charging hopper. After transferring the $FeSO_4$ filtrate from step (2) into a 250ml three-necked flask, add the $(NH_4)_2SO_4$ solution dropwise under stirring at 70℃. After this keep heating for 5 minutes to make sure that the reaction is completed. Then the solution is put into an evaporating dish and heated until a layer of tiny crystals can be observed on the surface of solution. Cool the concentrated solution to room temperature, filter it through a Buchner funnel and wash the crystals once with 5ml of 95% ethanol. The resulting crystals are weighed and calculate the yield.

2. Product test

(1) Analysis dose limited of Fe^{3+}

Dissolve 1g of product with 15ml of oxygen-free purified water in a 50ml beaker. After being transferred to a 25ml colorimetric tube, add 2ml 3mol/L HCl and 1ml 25% KSCN, continue to add oxygen-free purified water to the scale of 25ml, shake well. Compare the color with the series of standard samples to determine the purity grade of the product.

(2) Preparation of the standard samples

Separately take 0.01mol/L Fe^{3+} solution 5ml, 10ml, 20ml, and the following procedure is the same as test (1).

Grade I : 0.05mg (the content of Fe^{3+} in 1g product is less than 0.05mg)

Grade II : 0.10mg (the content of Fe^{3+} in 1g product is between 0.05mg and 0.10mg)

Grade III : 0.20mg (the content of Fe^{3+} in 1g product is between 0.10mg and 0.20mg)

3. Content assay

1.5g (accurate to 0.0001g) of product is dissolved in 100ml oxygen-free purified water. Then add 5ml of H_2SO_4, 2ml of H_3PO_4 and 200ml of oxygen-free purified water, shake well, and is titrated with $KMnO_4$ standard solution ($c = 0.02$mol/L) to pink which can keep 30 seconds.

The mass fraction ω, expressed as a "%" of $(NH_4)_2SO_4 \cdot FeSO_4 \cdot 6H_2O$, which can be calculated as follows:

$$\omega = c \cdot V \cdot M/1000m \times 100$$

Where c is the concentration of $KMnO_4$ standard titration solution (mol/L), V is the volume of $KMnO_4$ standard titration solution (ml), M is the molar mass of $FeSO_4 \cdot (NH_4)_2SO_4 \cdot 6H_2O$ (g/mol), m is the weight of the sample (g).

Appendix

Table 20-1 The solubility of ammonium sulfate at different temperatures

Temperature (℃)	Solubility (g/100g water)	Temperature (℃)	Solubility (g/100g water)
1	70.6	50	—
10	73	60	88
20	75.4	70	89.6
30	78	80	95.3
40	81	100	103.3

Table 20-2 The solubility of ferrous sulfate at different temperatures

Temperature(℃)	Solubility(g/100g water)	Temperature(℃)	Solubility(g/100g water)
1	15.6	57	—
10	20.5	65	—
20	26.5	70	50.9
30	32.9	80	43.6
40	40.2	90	37.3
50	48.6		

Table 20 – 3　The solubility of ferrous ammonium sulphate at different temperatures

Temperature (℃)	Solubility (g/100g water)	Temperature (℃)	Solubility (g/100g water)
1	12.5	40	33
10	—	50	40
20	—	60	—
30	—		

Vocabulary

ferrous	亚铁的	sulfate	硫酸盐
dilute	稀释的	grease stain	油污
Buchner funnel	布氏漏斗	electronic balance	电子天平
aqueous	水的	yield	收率
three-necked flask	三颈瓶	oxygen-free	无氧的
tilt-pour	倾倒	oxidize	氧化
purified water	纯化水	erlenmeyer flask	锥形瓶
filter	过滤	ethanol	乙醇

（编写：丁怀伟，英文核审：肖琰）

实验二十一　药用氢氧化铝的制备、鉴别、制酸力检查及含量测定

【预习内容】

1. 列举几种可生成 Al(OH)$_3$ 的方法。
2. 掌握沉淀的减压抽滤及其洗涤操作。

【思考题】

1. 为何反应温度在 50℃ 左右进行？
2. 如何检查制得的 Al(OH)$_3$ 有无 SO_4^{2-}？

【实验原理】

Al$_2$(SO$_4$)$_3$ 溶液和 Na$_2$CO$_3$ 溶液在加热条件下生成 Al(OH)$_3$：

$$Al_2(SO_4)_3 + 3Na_2CO_3 + 3H_2O = 2Al(OH)_3\downarrow + 3Na_2SO_4 + 3CO_2\uparrow$$

Al(OH)$_3$ 作为药用常用于治疗胃酸过多而合并的反酸等症状，适用于胃及十二指肠溃疡等疾病。

【仪器与试剂】

三颈瓶（250ml），量筒（10ml、100ml），具塞锥形瓶（250ml），量瓶（250ml），移液管（50ml），恒温水浴，研钵，抽滤装置，恒温烘箱，电子天平。

Na₂CO₃(2mol/L)，Al₂(SO₄)₃(s)，pH 试纸，HCl（滴定液，0.1mol/L），NaOH（试液），NaOH（滴定液，0.1mol/L），氨（试液），茜素磺酸钠（指示液），溴酚蓝（指示液），二甲酚橙（指示液），EDTA-2Na（滴定液，0.05mol/L），锌（滴定液，0.05mol/L），HAc-NH₄Ac（缓冲溶液，pH=6.0）。

【操作步骤】

1. Al(OH)₃的制备 称取13.32g的Al₂(SO₄)₃·18H₂O置于250ml三颈瓶中，加入20ml水，在50℃的水浴上搅拌使之溶解，然后滴加2mol/L Na₂CO₃溶液约35ml，控制pH 6.8~7.5，加料毕，继续搅拌20分钟，冷却至室温，抽滤，用纯化水洗涤3次，105℃~110℃下干燥6小时，粉碎得氢氧化铝，称重，计算收率。

2. 鉴别 取本品约0.5g，加稀盐酸10ml，加热使其溶解，作为供试品溶液。

（1）取供试品溶液，加 NaOH 试液，即生成白色胶状沉淀；分离，沉淀能在过量的 NaOH 溶液中溶解。

（2）取供试品溶液，加氨试液至生成白色胶状沉淀，滴加茜素磺酸钠指示液数滴，沉淀即显樱红色。

3. 检查 制酸力：取本品约0.12g，精密称定，置于250ml的具塞锥形瓶中，精密加入0.1mol/L的盐酸溶液50ml，密塞，在37℃不断振摇1小时，放冷后，加溴酚蓝指示液6~8滴，用 NaOH（0.1mol/L）滴定。每1g消耗0.1mol/L的盐酸液不得少于250ml。

4. 含量测定 取本品约0.6g，精密称定，加盐酸与水各10ml，加热溶解后，放冷至室温，滤过，滤液置于250ml量瓶中，滤器用水洗涤，洗液并入量瓶中，用水稀释至刻度，摇匀；精密量取25ml，加氨试液中和至恰析出沉淀，再滴加稀盐酸至沉淀恰溶解为止，加醋酸-醋酸铵缓冲液（pH=6.0）10ml，再精密加0.05mol/L的EDTA-2Na滴定液25ml，煮沸5分钟，放冷至室温，加二甲酚橙指示液1ml，用0.05mol/L的锌液滴定，至溶液自黄色转变为红色，并将滴定的结果用空白试验校正。每1ml的0.05mol/L的EDTA-2Na液相当于3.900mg的Al(OH)₃。

Experiment 21 Preparation, Identification, Neutralizing Capacity Examination and Content Assay of Medicinal Aluminum Hydroxide

Preview

1. List several methods which can produce Al(OH)₃.
2. Master the negative pressure filtration and washing operation.

Questions

1. Why is the reaction temperature kept at about 50℃?
2. How to check the existence of SO_4^{2-} in the prepared Al(OH)₃?

Principles

$Al_2(SO_4)_3$ reacts with Na_2CO_3 in aqueous solution to get aluminium hydroxide under heating.

$$Al_2(SO_4)_3 + 3Na_2CO_3 + 3H_2O = 2Al(OH)_3 \downarrow + 3Na_2SO_4 + 3CO_2 \uparrow$$

$Al(OH)_3$ as a drug commonly used to treat hyperacidity and combined acid reflux symptoms, suitable for gastric and duodenal ulcers and other diseases.

Instrument and Chemicals

three-necked flask (250ml), measuring cylinder(10ml, 100ml), conical flask with stopper (250ml), volumetric flask(250ml), volumetric pipet (50ml), thermostatic water bath, mortar, suction filter device, constant temperature oven, electronic balance.

Na_2CO_3 (2mol/L), $Al_2(SO_4)_3$ (s), pH test strips, HCl (VS, 0.1mol/L), NaOH (TS), NaOH (VS, 0.1mol/L), ammonia (TS), sodium alizarin sulfonate (IS), bromophenol blue (IS), xylenol orange (IS), EDTA-2Na (VS, 0.05mol/L), zinc (VS, 0.05mol/L), HAc-NH_4Ac (BS, pH = 6.0).

Procedures

1. Preparation

13.32g $Al_2(SO_4)_3 \cdot 18H_2O$ and 20ml water are put into a 250ml three-necked flask which is heated at 50℃ in the water bath until $Al_2(SO_4)_3$ is dissolved. Then add dropwise 35ml of 2mol/L Na_2CO_3 solution under stirring and control pH between 6.8 and 7.5. Continue stirring for 20 minutes after completion of addition of Na_2CO_3. Cool to room temperature and filter under vaccum and wash the crystals three times with purified water. After filtering, drying for 6 hours at 105℃ to 110℃, smashing, the product is obtianed. The resulting crystal is weighed and calculate the yield.

2. Identification

Heat and dissolve about 0.5g $Al(OH)_3$ in 10ml of dilute hydrochloric acid. A solution of product being examined is obtained.

(1) Add NaOH TS to a solution of the substance being examined; a gelatinous white preciptate appears, which is soluble in an excess of NaOH TS.

(2) Add ammonia TS to a solution of the substance being examined until a gelatinous white preciptate is formed, add a few drops of sodium alizarin sulfonate IS; the preciptate becomes cherry red.

3. Neutralizing capacity examination

To about 0.12g $Al(OH)_3$, accurately weighed, in a 250ml conical flask with stopper, add 50ml of 0.1mol/L HCl VS, accurately measured, stopper the flask and shake for 1 hour continuonsly at 37℃. Allow the solution to cool, add 6 drops to 8 drops of bromophenol blue IS, and titrate with 0.1mol/L NaOH VS. The volume of 0.1mol/L HCl VS consumed is not less than 250ml for each gram of the substance being examined.

4. Content Assay

Dissolve by heat about 0.6g accurately weighed, in 10ml each of HCl and water, cool to room temperature and filter. Transfer the filrate to a 250ml volumetric flask, wash the filter cake with water, combine the washings to the flask, dilute with water to volume and mix well. Measure accurately 25ml of the resulting solution, neutralize with ammonia TS until the precipitate is just dissloved, add 10ml of HAc-NH_4Ac BS (pH 6.0) and then 25ml of 0.05mol/L EDTA-2Na VS, accurately measured. Boil the solution for 5 minutes, cool it to room temperature, add 1ml of xylenol orange IS, titrate with 0.05mol/L zinc VS until the solution changes from yellow to red. Perform a blank determination and make any necessary correction. Each ml of 0.05mol/L EDTA-2Na VS is equivalent to 3.900mg of Al(OH)$_3$.

Vocabulary

aluminum hydroxide	氢氧化铝	alluminium sulphate	硫酸铝
sodium carbonate	碳酸钠	three-necked flask	三颈瓶
measuring cylinder	量筒	conical flask with stopper	具塞锥形瓶
pH test strips	pH 试纸	bromophenol blue	溴酚蓝
test solution (TS)	试液	volumetric solutions (VS)	滴定液
indicator solution (IS)	指示液	buffer solution (BS)	缓冲溶液

（编写：丁怀伟，英文核审：肖琰）

实验二十二 药用碱式碳酸铋的制备、鉴别、制酸力检查及含量测定

【预习内容】

1. 药用碱式碳酸铋的制备原理。
2. 药用碱式碳酸铋制酸力的测定原因。

【实验原理】

碱式碳酸铋是一种组成不定的碱式盐。按干燥品计算，含铋（Bi）应为 80.0%～82.5%。它是白色或微黄色粉末，不溶于水或乙醇，可用作抗酸药和收敛药。可采用硝酸铋在碳酸钠溶液中的水解反应进行制备，反应式如下：

$$Bi(NO_3)_3 \xrightarrow{Na_2CO_3, H_2O, 50℃\sim55℃, pH\ 8.5\sim9} (BiO)_2CO_3 \cdot \frac{1}{2}H_2O$$

【实验仪器与药品】

托盘天平，量筒（50ml，10ml），烧杯（50ml×2），三颈瓶（100ml），控温磁力搅拌器，布氏漏斗，吸滤瓶，电烘箱，60 目筛，锥形瓶（250ml），具塞锥形瓶（250ml），电子天平，X 射线粉末衍射仪。

Bi$(NO_3)_3 \cdot 5H_2O$（s），Na_2CO_3（s），溴酚蓝（指示液），二甲酚橙（指示液），HNO_3（3→10），HNO_3，HCl（滴定液，0.1mol/L），NaOH（滴定液，0.1mol/L），EDTA-2Na（滴定液，0.05mol/L），稀HCl，Na_2S（试液），硫脲溶液（10%），KI（试液）。

【操作步骤】

1. 制备 称取10g固体Bi$(NO_3)_3 \cdot 5H_2O$，置于50ml烧杯中，加20ml水，搅拌至晶体颗粒全部变成乳浆；称取5g Na_2CO_3，置于100ml三颈瓶中，加20ml水，用水浴加热溶解；在搅拌下，将乳浆快速加入至50℃~55℃的Na_2CO_3溶液中，调节pH值至8.5~9.0，继续搅拌反应1小时。反应停止后，减压抽滤，用40℃~45℃的水洗涤滤饼至中性；将滤饼放入烘箱中，在60℃~70℃下干燥；当近干时，再90℃~100℃下干燥1小时；将其碾碎，过60目筛，得产品。

2. 结构表征 X线粉末衍射方法可用于该产品的结构表征，如果测试结果与碱式碳酸铋的标准图谱一致，表明所得产品为碱式碳酸铋。

3. 鉴别

（1）取约0.2g本品，加2ml稀HCl，即发生泡沸并溶解。溶液分成两等份：一份中加水稀释，即生成白色沉淀，再加Na_2S试液，沉淀变为棕褐色；另一份中加10%硫脲溶液1ml，即显深黄色。

（2）取本品约50mg，加1ml的HNO_3溶解后，加水10ml；分取2ml，滴加KI试液，即生成棕黑色沉淀，再加过量的KI试液，沉淀即溶解成黄橙色的溶液。

4. 干燥失重 取本品，在105℃干燥至恒重，减失重量不得过1.0%。

5. 制酸力 取本品约0.50g，精密称定，置具塞锥形瓶中，精密加入0.1mol/L的HCl滴定液50ml，密塞，在37℃不断振摇1小时，放冷，加水50ml，加溴酚蓝指示液8滴，用0.1mol/L的NaOH滴定液滴定剩余的盐酸。按干燥品计算，每1g消耗0.1mol/L的HCl滴定液不得少于38ml。

6. 含量测定 取本品约0.2g，精密称定，加HNO_3溶液（3→10）5ml使溶解，再加水100ml和二甲酚橙指示液3滴，用0.05mol/L EDTA-2Na滴定液滴定至淡黄色。每1ml的0.05mol/L EDTA-2Na滴定液相当于10.45mg的Bi。

附注

表22-1 碱式碳酸铋的标准X线粉末衍射数据

d (Å)	I/I_0	d (Å)	I/I_0
6.86	20	1.86	5
3.72	40	1.81	2
3.42	14	1.75	20
2.95	100	1.72	11
2.73	40	1.68	6
2.54	7	1.62	23

续表

d (Å)	I/I_0	d (Å)	I/I_0
2.28	10	1.47	7
2.24	3	1.46	2
2.14	22	1.42	5
1.93	20	1.38	5

Experiment 22　Preparation, Identification, Neutralizing Capacity Examination and Content Assay of Medicinal Bismuth Subcarbonate

Preview

1. The principle of preparation of medicinal bismuth subcarbonate.
2. The cause of determining neutralizing capacity to acid for medicinal bismuth subcarbonate.

Principle

Bismuth subcarbonate is a basic salt with varying composition. It contains not less than 80.0% and not more than 82.5% of Bi, calculated on the dried basis. It is white or slightly pale yellow powder and insoluble in water or ethanol, which can be served as antacid and astringent. It can be obtained by hydrolyzing bismuth nitrate in sodium carbonate solution, which can be indicated by the following equation.

$$Bi(NO_3)_3 \xrightarrow{Na_2CO_3, H_2O, 50℃\sim55℃, pH\ 8.5-9} (BiO)_2CO_3 \cdot \frac{1}{2}H_2O$$

Instruments and Chemicals

platform balance, graduated cylinder (50ml, 10ml), beaker (50ml × 2), electric three-necked flask (100ml), thermostatic magnetic stirrer, Buchner funnel, suction flask, oven, 60-mesh sieve, conical flask (250ml), conical flask with stopper (250ml), electronic balance, X-ray powder diffraction apparatus.

$Bi(NO_3)_3 \cdot 5H_2O$ (s), Na_2CO_3 (s), bromophenol blue indicator solution, xylenol orange indicator solution, HNO_3 (3→10), HNO_3, HCl (0.1mol/L) volumetric solution, NaOH (0.1mol/L) volumetric solution, EDTA-2Na (0.05mol/L) volumetric solution, dilute HCl, Na_2S test solution, thiourea solution (10%), KI test solution.

Procedures

1. Preparation

Mix 10g of $Bi(NO_3)_3 \cdot 5H_2O$ with 20ml of water in a 50ml beaker, then stir continuously until all the crystals turn to emulsion. Put 5g of Na_2CO_3 into a 100ml three-necked flask, then add 20ml of water and warm on a water bath to dissolve it. With stirring, transfer the emulsion rapidly to Na_2CO_3 solution at 50℃—55℃, adjust pH to 8.5—9.0, then keep stirring for 1 hour. Filter it with suction pump and wash filter cake to be neutral with warm water at 40℃—45℃ after reaction. Dry the filter cake in oven at 60℃—70℃. Then heat up to 90℃—100℃ when it is closed to dried, and keep the temperature for 1 hour. After crushing and pouring it through 60-mesh sieve, the product can be obtained.

2. Structural characterization

The structure of the product is characterized by X-ray powder diffraction. If the test result is consistent with the standard X-ray powder diffraction date of bismuth carbonate, indicating that the obtained product can be identified as bismuth carbonate.

3. Identification

(1) To about 0.2g of product, add 2ml of dilute HCl and dissolve it with effervescence. Divide this solution into two equal portions. Dilute one portion with water, a white precipitate is produced which changes to brown color on adding Na_2S test solution. Add 1ml of 10% thiourea solution to the other, a deep yellow color appears.

(2) Dissolve about 50mg of product in 1ml of HNO_3 and add 10ml of water. To 2ml of this solution, add KI test solution dropwise. A brownish-black precipitate is produced, which is dissolved in an excess of KI test solution to form a yellowish-orange solution.

4. Loss on drying

When dried to constant weight at 105℃, the product loses not more than 1.0% of its weight.

5. Neutralizing capacity

To about 0.50g of product, accurately weighed, in a conical flask with stopper, add 50ml of 0.1mol/L HCl volumetric solution accurately measured, stopper the flask and shake for 1 hour continuously at 37℃. Cool the solution, and add 50ml of water and 8 drops of bromophenol blue indicator solution, then titrate the residual HCl with 0.1mol/L NaOH volumetric solution. The volume of 0.1mol/L HCl volumetric solution consumed is not less than 38ml for each gram of product, calculated on the dried basis.

6. Assay

Dissolve about 0.2g of product, accurately weighed, in 5ml of HNO_3 (3→10), add 100ml of water and 3 drops of xylenol orange indicator solution, titrate with 0.05mol/L EDTA-2Na volumetric solution to pale yellow color. Each ml of EDTA-2Na volumetric solution is equivalent to 10.45mg of Bi.

Appendix

Table 22-1 Standard X-ray powder diffraction data of bismuth carbonate

$d(Å)$	I/I_0	$d(Å)$	I/I_0
6.86	20	1.86	5
3.72	40	1.81	2
3.42	14	1.75	20
2.95	100	1.72	11

续表

$d(\text{Å})$	I/I_0	$d(\text{Å})$	I/I_0
2.73	40	1.68	6
2.54	7	1.62	23
2.28	10	1.47	7
2.24	3	1.46	2
2.14	22	1.42	5
1.93	20	1.38	5

Vocabulary

medicinal	药用的	bismuth	铋
subcarbonate	碱式碳酸盐	examination	检查
composition	组成	powder	粉末
insoluble	不溶解的	antacid	抗酸药
astringent	收敛药	neutralizing capacity	制酸力
calculated on the dried basis	按照干燥品计	bismuth nitrate	硝酸铋
sodium carbonate	碳酸钠	electronic balance	电子天平
platform balance	托盘天平	graduated cylinder	量筒
three-neckedflask	三颈瓶	Buchner funnel	布氏漏斗
thermostaticmagnetic stirrer	恒温磁力搅拌	electric oven	烘箱
suction flask	吸滤瓶	60-mesh sieve	60目筛
bromophenol blue	溴酚蓝	indicator solution	指示液
xylenol orange	二甲酚橙	loss on drying	干燥失重
volumetric solution	滴定液	thiourea	硫脲
emulsion	乳液	suction pump	抽滤泵
structural characterization	结构表征	filter cake	滤饼
X-ray powder diffraction	X线粉末衍射	effervescence	泡腾
stopper (n. ,vt.)	塞子,用塞子塞住	conical flask	锥形瓶
titrate (vt. ,vi.)	滴定		

（编写：刘迎春，英文核审：夏丹丹）

第六章　前沿性实验和综合设计实验

Chapter 6　Frontier Experiments and Comprehensive Designing Experiments

实验二十三　二氧化钛纳米粒子的制备及其光催化活性的测定

【预习内容】
1. 二氧化钛纳米粒子的制备方法。
2. 二氧化钛纳米粒子的应用。

【思考题】
1. 溶胶-凝胶法制备纳米材料的影响因素？
2. 如何根据X射线衍射结果，求算纳米晶尺寸？
3. 取用钛酸丁酯时应注意哪些问题？

【实验原理】
溶胶-凝胶法是在温和条件下制备无机化合物或无机材料的重要方法。简单的说，溶胶-凝胶法就是用含高化学活性组分的化合物作前驱体，在液相下将这些原料均匀混合，并进行水解、缩合化学反应，在溶液中形成稳定的透明溶胶体系，溶胶经陈化胶粒间缓慢聚合，形成三维空间网络结构的凝胶，凝胶网络间充满了失去流动性的溶剂，形成凝胶。凝胶经过干燥、烧结固化制备出分子乃至纳米亚结构的材料。根据现阶段研究表明，适当地控制溶液pH、溶液浓度、反应温度和反应时间可以制备出小至纳米的超细粉体。

纳米二氧化钛具有极大的体积效应、表面效应、光学特性、颜色效应，故在光、电及催化环保等方面显示出其特殊性质。溶胶-凝胶法制备TiO_2具有合成温度低、纯度高、均匀性好、化学成分准确、工艺简单等特点。本实验以钛酸丁酯为原料，用溶胶-凝胶法制备TiO_2纳米粒子，并进行XRD、IR、TG/DTA和光催化活性的表征。

【仪器与试剂】
721型分光光度计，热分析天平，超声波清洗机KQ-5200DB，X射线衍射仪，真空干燥箱DZF-6020，马弗炉。

无水乙醇（99%），冰乙酸（99%），浓盐酸（12mol/L），钛酸丁酯（97%），亚甲基蓝（96%）。

【操作步骤】
1. 将6ml的无水乙醇、2ml乙酸、1.5ml浓盐酸和3ml纯化水混合均匀，配成溶液

A；然后将24ml无水乙醇、17ml钛酸丁酯混合搅拌均匀，制成溶液B；在磁力搅拌下将A液向B液中慢慢滴加，滴加时间约为20分钟，然后继续搅拌2小时，得到湿胶。将制得的湿胶静置陈化10小时，再放入烘箱于60℃条件下，干燥12小时，得到凝胶。将得到的粉于马弗炉中500℃条件下，煅烧2小时，即得白色TiO_2纳米粒子。准备粉末样品，做IR及TG/DTA表征。

2. 光催化降解测试过程如下：取0.02g制备好的TiO_2光催化剂超声分散到60ml浓度为10mg/L的亚甲基蓝水溶液中。在黑暗处超声分散1小时以确保达到吸附平衡，再在自制功率为30W的紫外光催化反应器中恒温磁力搅拌，每30分钟取5ml溶液，高速离心分离10分钟后取上层清液，用分光光度计测定亚甲基蓝在664nm处的吸光度，与降解前亚甲基蓝溶液的吸光度进行比较，确定亚甲基蓝的降解率。

$$降解率 = (A_0 - A)/A_0 \times 100\% \qquad (24-1)$$

A_0为吸附脱附平衡后，亚甲基蓝在664nm处的初始吸光度；A为光照30分钟后，亚甲基蓝在664nm处的吸光度。

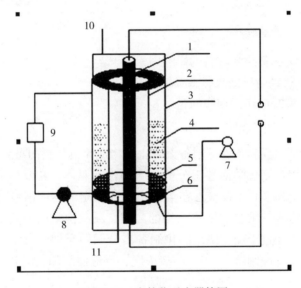

图23-1 光催化反应器简图

1. 紫外灯管（距离石英管内表面5mm）；2. 石英玻璃管（2mm）；3. 有机玻璃管；
4. 主反应区（间隙宽度5mm）；5. 气体分布器（管式）；6. 法兰底盘；7. 气体风机；
8. 液体循环泵；9. 配液池；10. 出气口；11. 采样口或出液口

Experiment 23 Preparation of Nano TiO_2 Particles and Determination of Its Photocatalytic Activity

Preview

1. Preparation of titanium dioxide nanoparticles.

2. Application of titanium dioxide nanoparticles.

Questions

1. What are the effect factors on preparation of nanomaterials by sel-gel method?

2. How to calculate the size of nanocrystalline according to the results of X-ray diffraction?

3. What should be taken notice of when taking use of tetrabutyl titanate?

Principles

Sol-Gel is an important method on preparing inorganic chemicals and inorganic materials under mild conditions. In short, mix together the high chemically reactive compounds acting as precursor and then perform hydrolysis and condensation reaction in liquid phase until a stable and transparent colloid system is formed. After the slow polymerization of the colloidal particles, 3D space network is established. When the gel network full of the fluidized solvent, the colloid system further forms a gel. After the process of drying and sintering, molecular with nano substructure materials are prepared. At present, studies show that nanoscale ultrafine powder can be synthesized by controlling of pH, concentration, reaction temperature and time.

Nano TiO_2 material, because of its volume effect, surface effect, colour effect and optical properties, exhibits special applications on light, electricity, catalysis and environmental protection. As the preparation of TiO_2 particles by Sol-Gel method shows the advantage of high-purity, uniformity, accurant component and simply technology, Sol-Gel method is chosen to prepare nano TiO_2 particles in this experiment using tetrabutyl titanate as material. At last the nano TiO_2 particles are characterized by XRD, IR, TG/DTA and photocatalytic activity.

Instruments and Chemicals

721-visible spectrophotometer, thermo-analytical balance, ultrasonic cleaner, X-ray diffraction, vacuum drying oven, muffle furnace.

absolute alcohol(99%), acetic acid(99%), hydrochloric acid(12mol/L), tetrabutyl titanate(97%), methylene blue(96%).

Procedures

1. Firstly, mix 6ml absolute ethyl alcohol, 2ml acetic acid, 1.5ml hydrochloric acid and 3ml purified water together, forming solution A. Secondly, mix 24ml absolute ethyl alcohol and 17ml tetrabutyl titanate, forming solution B. Then drop solution A to B slowly within 20min, under magnetic stirring, keep the reaction for 2h to form wet colloids. After room-temperature aging for 10h, drying for 12h at 60℃, gel forms. Place the gel into the muffle to carry on burning for 2h at 500℃ then white nano TiO_2 particles are obtained.

2. The procedure of photocatalytic degradation is as follows: disperse 0.02g TiO_2 photo catalytic into 60ml of 10mg/L methylene blue by ultrasonic method. Keep being ultrasonic in dark for 1h to get the adsorption equilibrium. Continue the degradation in the self-made UV-photocatalytic reactor with the power 30W under magnetic stirring. Remove 5ml of the methylene blue solution every 30 minutes and then measure the absorbance of methylene blue samples

separated by high speed centrifugation, at 664nm. Compare the absorbance difference of the methylene blue samples before and after depredating, the degradation rate is determined.

$$\text{degradation rate} = (A_0 - A)/A_0 \times 100\% \tag{24-1}$$

A_0 is the initial absorbance of methylene blue samples at 664nm. A is the absorbance of methylene blue samples under the condition of illumination for 30 minutes at 664nm.

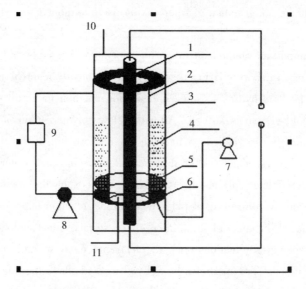

Fig. 23-1 Photocatalytic reactor-schematic

1. UV lamp tube (5mm distance from the inside surface); 2. quartz glass tube (2mm); 3. plexiglass tube; 4. main reacted region (gap width 5mm); 5. gas distributor (tube style); 6. forging-die; 7. air device; 8. liquid circulating pump; 9. distributor; 10. outlet; 11. sampling port or export

Vocabulary

sel-gel method	溶胶-凝胶法	hydrolysis	水解
condensation react	缩合反应	tetrabutyl titanate	钛酸丁酯
methylene blue	亚甲基蓝	XRD	X射线衍射
IR	红外光谱	TG	热重法
DTA	差热分析法	nanocrystalline	纳米晶

（编写：张莹，英文核审：肖琰）

实验二十四 纳米四氧化三铁的化学共沉淀法制备及表征

【预习内容】

1. 不同氧化态铁盐的物理化学性质。
2. 纳米氧化铁在医学领域的应用。

【思考题】
1. 比较纳米四氧化三铁不同制备方法的优缺点。
2. 化学共沉淀法制备纳米四氧化三铁的关键影响因素有哪些？都是如何影响的？

【实验原理】
共沉淀法是目前最普遍使用的方法，它是按方程式：
$$Fe^{2+} + 2Fe^{3+} + 8OH^- \rightarrow Fe_3O_4 + 4H_2O$$
这种方法又可细分为 Massart 水解法和滴定水解法。其中 Massart 水解法是将 Fe^{2+} 和 Fe^{3+} 混合液直接加入强碱性水溶液中，铁盐瞬间水解、结晶，形成 Fe_3O_4 晶体。滴定水解法是将稀释碱溶液滴加到 Fe^{2+} 和 Fe^{3+} 的混合液中，随 pH 增高，达到 6~7 时水解生成 Fe_3O_4 晶体。从这两种水解过程可以看出，Fe_3O_4 纳米晶的生成机制是不同的，Massart 水解法采用快速成核的方法，更有利于生成均一尺寸的 Fe_3O_4 纳米晶。通常是把 Fe^{3+} 和 Fe^{2+} 盐溶液以 2∶1（或更大）的比例混合，在一定温度和 pH 值下加入过量（2~3 倍）的 NaOH，高速搅拌进行沉淀反应，然后将其沉淀洗涤、过滤、干燥、烘干，制得尺寸为 8~10 nm 的纳米 Fe_3O_4 微粒。

【仪器与试剂】
X－射线衍射仪，三颈瓶（250ml），量筒，氮气袋，滴液管，永磁铁，磁力搅拌器，pHS－25 型数显酸度计，真空干燥箱。

硫酸铁 $[Fe_2(SO_4)_3 \cdot xH_2O]$，硫酸亚铁（$FeSO_4 \cdot 6H_2O$），NaOH（3mol/L），$NH_3 \cdot H_2O$（28%），$NH_4Cl$（固体）。

【操作步骤】
1. 取 pH 9 的 $NH_3 - NH_4Cl$ 缓冲溶液 15ml，置于 250ml 三颈瓶中，N_2 保护。取 0.5mol/L 的 $Fe_2(SO_4)_3$ 溶液 10ml 和 0.5mol/L $FeSO_4$ 溶液 10ml，混合均匀后置于 50ml 滴液管中。另取 3mol/L NaOH 溶液 14ml 置于另一 50ml 的滴液管中。水浴升温至 50℃ 后，磁力搅拌条件下，滴加 Fe^{2+} 和 Fe^{3+} 的混合液，同时滴加 NaOH 溶液，控制滴加速度，15 分钟滴加完毕。继续反应 10 分钟，结束反应。

2. 取出反应液，转移至烧杯，冷却，磁沉降，待液体澄清，沉淀均被吸附在烧杯底部，弃去上清液。重复洗涤沉淀数次，再将沉淀放入真空干燥箱 60℃ 条件下干燥。研磨，即得纳米氧化铁。

3. 取产物在 X 射线衍射仪上测定物相。采用 Cu（=0.15406nm）靶，石墨单色器，工作电压 40kV，工作电流 30mA，扫描速率 6°/min，扫描范围 20°~80°。将产物纳米粒子的 X 射线衍射谱与 X 射线衍射标准卡片（PDF 卡片）的衍射峰相对照，确定 Fe_3O_4 纳米晶的生成。

Experiment 24 Preparation and Characterization of Nano Fe_3O_4 by Chemical Coprecipitation

Preview

1. Physicochemical properties of different iron oxide.
2. Application of iron oxide nanomaterial in the medical field.

Questions

1. Compare the advantages and disadvantages of different preparation methods of Nano Fe_3O_4.
2. The key influencing factors of the preparation of nanometer Fe_3O_4 by chemical coprecipitation. How does it affect?

Principles

Coprecipitation is the most popular method at the moment, according to the equation:

$$Fe^{2+} + 2Fe^{3+} + 8OH^- \rightarrow Fe_3O_4 + 4H_2O$$

This method can be divided into Massart hydrolysis and titration hydrolysis. Massart hydrolysis is to directly add the mixture of Fe^{2+} and Fe^{3+} to strong alkaline solutions, ferric salt instantly hydrolyze and crystalize, and subsequently crystal Fe_3O_4 is formed. Titration hydrolysis is to add the diluted alkali solution to the mixture of Fe^{2+} and Fe^{3+}, with pH increasing to 6—7, crystal Fe_3O_4 is formed by hydrolysis. From the process of these two hydrolyses, it can be found that the generative mechanism of the Fe_3O_4 nanocrystalline is different. Massart hydrolysis utilizes the method of rapid nucleation, which is more conducive to forming the mono size Fe_3O_4 nanocrystalline. Generally, mix Fe^{3+} and Fe^{2+} in the portion of 2:1 (or higher); under a certain temperature and pH value, add excessive (2—3times) NaOH, stir at high speed for precipitation reaction; and then wash, filter, dry, make the nanometer Fe_3O_4 particles in size of 8—10nm.

Instruments and Chemicals

X-ray diffractometer, three-necked flask, measuring cylinder, nitrogen bag, dropping pipettes, permanent magnet, magnetic stirrer, pHS-25 acid meter, vacuum drying oven.

ferric sulfate hydrate ($Fe_2(SO_4)_3$) · xH_2O, iron sulfate ($FeSO_4$ · $6H_2O$), sodium hydroxide (NaOH), ammonia (NH_3 · H_2O), ammonium chloride NH_4Cl.

Procedures

1. Take out 15ml buffer solution of NH_3-NH_4Cl, pour into 250ml three-necked flask, N_2 protect. Take out 10ml of 0.5mol/L $Fe_2(SO_4)_3$ and 10ml of 0.5mol/L $FeSO_4$, mix them and place them into a 50ml dropping pipettes. Take out 3mol/L NaOH 14ml and place it into another 50ml dropping pipettes. After the temperature of water bath increasing to 50℃, under magnetic stirring, dropwise add the mixture of Fe^{2+} and Fe^{3+}, meanwhile add NaOH solution,

control dropping speed, addition is completed within 15min. Continue to react for another 10min and end the reaction.

2. Take out the reaction liquid of the sample, and transfer it into a beaker. After cooling, magnetic settlement, the liquid gets clarified. When precipitate is absorbed on the bottom of the beaker, remove all supernate. Wash the precipitate over and over again, and then place it into the vacuum drying oven and dry at 60oC. After grinding, iron oxide particles are formed.

3. Take the product to measure phase on the X-ray diffractometer. Use Cu target, graphite monochromator, working voltage of 40kV, working current of 30mA, scanning rate in 6°/min, range of scanning from 20°—80°. Compare the X-ray diffraction patterns of the nano-particle to the diffraction maximum of the PDF standard card, and determine the formation of nano Fe_3O_4 particles.

Vocabulary

chemical coprecipitation	化学共沉淀法	nanocrystalline	纳米晶
dropping pipettes	滴液管	three-necked flask	三颈瓶
ferric sulfate	$Fe_2(SO_4)_3$	iron sulfate	$FeSO_4$

（编写：张莹，英文核审：肖琰）

实验二十五 十二钨硅酸的制备、萃取分离及表征

【预习内容】

1. 十二钨硅酸的性质。
2. 萃取分离操作。
3. 了解用红外光谱、紫外吸收光谱及热分析法对产物进行表征的方法。

【思考题】

1. 钨硅酸有哪些性质？
2. 在 $[SiW_{12}O_{40}]^{4-}$ 离子中有几种不同结构的氧原子？每种结构氧原子各有多少个？
3. 十二钨硅酸易被还原，在制备过程中要注意哪些问题？

【实验原理】

钒、铌、钼、钨等元素易形成同多酸和杂多酸。在碱性溶液中 W（Ⅵ）以钨酸根 WO_4^{2-} 形式存在，随着溶液 pH 的减小，WO_4^{2-} 逐渐聚合成多酸根离子（如下表所示）。

H^+/WO_4^{2-}（物质的量之比）	同多酸阴离子	
1.14	$[W_7O_{24}]^{6-}$	仲钨酸根（A）离子
1.17	$[W_{12}O_{42}H_2]^{10-}$	仲钨酸根（B）离子
1.50	$\alpha-[H_2W_{12}O_{40}]^{6-}$	钨酸根离子

H^+/WO_4^{2-}（物质的量之比）	同多酸阴离子	
1.60	$[W_{10}O_{32}]^{4-}$	十钨酸根离子
……	……	……

如在酸化过程中，加入一定量的硅酸盐，则可生成具有 Keggin 结构的十二钨硅酸 $[SiW_{12}O_{40}]^{4-}$。反应如下：

$$12\ WO_4^{2-} + SiO_3^{2-} + 22H^+ = [SiW_{12}O_{40}]^{4-} + 11\ H_2O$$

下面以十二钨硅阴离子为例介绍典型 Keggin 结构（图 25-1）。

$[SiW_{12}O_{40}]^{4-}$ 具有 Td 对称性。中心杂原子呈现 SiO_4 四面体配位。配原子呈 WO_6 八面体配位。3 个 WO_6 八面体为一组形成三金属簇 W_3O_{10}，它们共边相连，共有 4 组三金属簇。三金属簇之间以及与中心四面体之间都是共角相连。一共 12 个八面体围绕着中心四面体。

$H_4[SiW_{12}O_{40}] \cdot nH_2O$ 易溶于水及乙醚、丙酮中，在强碱性水溶液不稳定，而在酸性水溶液中较稳定。本实验利用钨硅酸在强酸性溶液中易与乙醚生成加合物而被乙醚萃取的性质来制备十二钨硅酸。

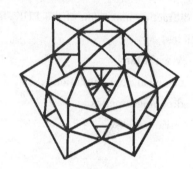

图 25-1 Keggin 结构图

钨硅酸高水合物，在空气中易风化也易潮解。对水合物晶体进行热分析，从热重（TG）曲线看出，水合物在 30℃～165℃ 及 165℃～310℃ 温度范围，有两个失水吸热峰。另外 DTA 曲线上，在 540℃ 附近出现 Keggin 结构被破坏后，由无序状态向有序结构转化的强吸热峰。十二钨硅酸不仅有强酸性，还有氧化还原性，在紫外光（260nm 附近）作用下，可以发生单电子或多电子还原反应。

【仪器与试剂】

差热天平，红外光谱仪，UV-240 型分光光度计，烧杯（100ml，250ml，50ml），磁力加热搅拌器，滴液漏斗（100ml），分液漏斗（250ml），蒸发皿，水浴锅，微型抽滤装置，表面皿，吸量管。

$Na_2WO_4 \cdot 2H_2O$(s)，$Na_2SiO_3 \cdot 9H_2O$(s)，HCl（6mol/L），乙醚，H_2O_2（3%，或溴水）。

【操作步骤】

1. 十二钨硅酸的制备

（1）十二钨硅酸溶液的制备　称取 25.0g $Na_2WO_4 \cdot 2H_2O$ 置于烧杯中，加入 50ml 水，再加入 1.9g $Na_2SiO_3 \cdot 9H_2O$，置于磁力加热搅拌器上加热搅拌，使其溶解。将混合物加热至近沸，由滴液漏斗以 1~2 滴/秒的速度加入约 10ml 浓盐酸，开始滴入浓 HCl 时，有黄色钨酸沉淀出现，要继续缓慢滴加至溶液 pH 为 2 时，保持 30 分钟左右。将混合物冷却。

（2）酸化、乙醚萃取十二钨硅酸　待上述溶液冷却后转移至分液漏斗中，加入乙醚（约为混合物液本体积的 1/2），分四次向其中加入 10ml 浓盐酸，充分振荡，萃取，静止后液体分三层，上层是溶有少量杂多酸的醚，中间是氯化钠、盐酸和其他物质的水溶液，下层是油状的杂多酸醚合物。将下一层醚合物分出，放于蒸发皿中，加水 4ml，水浴蒸发至溶液表面有晶体析出时为止，冷却结晶，抽滤，即可得到产品。

2. 测定产品热重（TG）曲线及差热分析（DTA）曲线　取少量样品，在热分析仪上，测定室温至 650℃ 范围内的 TG 曲线及 DTA 曲线。并计算样品的含水量，以确定水合物中结晶水数目。

3. 测定紫外吸收光谱　配制 5×10^{-5} mol/L 十二钨硅酸溶液，用 1cm 比色皿，以水为参比，在 UV-240 型分光光度计上，记录波长范围为 200～400nm 的吸收曲线。

4. 测定红外光谱　样品用 KBr 压片，在红外光谱仪上记录 400～4000cm^{-1} 范围的红外光谱图，并标识其主要的特征吸收峰。

Experiment 25　The preparation, extraction and characterization of dodecatungstosilicic acid

Preview

1. The properties of dodecatungstosilicic acid.
2. The operation of extraction for separation.
3. The characterization approach of IR, UV absorption spectra and thermogravimetric synthesis.

Questions

1. What are the properties of dodecatungstosilicic acid?
2. How many types of oxygen atoms with different coordination are there in the $[SiW_{12}O_{40}]^{4-}$? How many oxygen atoms are there in each type?
3. Dodecatungstosilicic acid is easily reduced, what problems we should pay attention to in the preparation process?

Principles

Vanadium, niobium, molybdenum, tungsten etc. easily form isopoly acids and heteropoly acids. In alkaline solution W(VI) exists as WO_4^{2-}, with the decrease of pH value, WO_4^{2-} gradually aggregate into isopoly ions (see table below).

H^+ / WO_4^{2-} (mole ratio)	Isoploy acid
1.14	$[W_7O_{24}]^{6-}$
1.17	$[W_{12}O_{42}H_2]^{10-}$
1.50	$\alpha-[H_2W_{12}O_{40}]^{6-}$
1.60	$[W_{10}O_{32}]^{4-}$
……	……

In the acidification process, $[SiW_{12}O_{40}]^{4-}$ with Keggin structure can be generated when a certain amount of silicate is added. The reaction is:

$$12\ WO_4^{2-} + SiO_3^{2-} + 22H^+ = [SiW_{12}O_{40}]^{4-} + 11\ H_2O$$

We introduced dodecatungstosilicic acid with typical Keggin structure as example.

$[SiW_{12}O_{40}]^{4-}$ exhibits Td symmetry. The central hetero atom presents SiO_4 tetrahedral. The ligand atoms present WO_6 octahedral. Three WO_6 octahedron form three metal cluster W_3O_{10} which are connected by edge sharing. There are four groups of three metal clusters, between them or with the center tetrahedral are connected by angle sharing. There are a total of 12 octahedrons around the center tetrahedron.

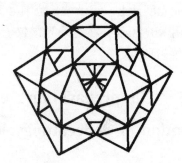

Fig. 25 – 1 Keggin structure

$H_4[SiW_{12}O_{40}] \cdot nH_2O$ is easy to be dissolved in water, ether or acetone, and unstable in strong alkaline aqueous solution, however it is more stable in acidic aqueous solution. In this experiment, dodecatungstosilicic acid is prepared by forming ether adduct and separated the ether extraction in strong acid solution.

Tungstosilicic acid hydrate is easily deliquescent in the air. The TG curve shows that hydrate crystals have two dehydration endothermic peak in temperature range of 30℃—165℃ and 165℃—310℃. DTA curve also shows that Keggin structure was destroyed corresponding to a strong endothermic peak in the vicinity of 540℃, where the disordered state turns to the ordered structure. Dodecatungstosilicic acid is not only a strong acid, but also oxidant, which can transfer one or more electrons reduction reaction under UV light (260nm nearby).

Instrument and Chemicals

Differential thermal balance, infrared spectroscopy, UV – 240 spectrophotometer, beaker (100ml, 250ml, 50ml), magnetic stirrer, dropping funnel (100ml), separating funnel (250ml), evaporating dish, water bath, micro-filtration device, a watch glass, pipette.

$Na_2WO_4 \cdot 2H_2O$ (s), $Na_2SiO_3 \cdot 9H_2O$ (s), HCl (6mol/L), diethyl ether, H_2O_2 (3%) or bromine water.

Procedures

1. Preparation of dodecatungstosilicic acid

(1) Add 25.0g $Na_2WO_4 \cdot 2H_2O$ to a beaker, add 50ml water, then add 1.9g $Na_2SiO_3 \cdot 9H_2O$, heating and stirring on a magnetic stirrer to dissolve. The mixture is heated to near boil-

ing, and then add 10ml concentrated hydrochloric acid at a velocity of 1—2day per second through dropping funnel. The yellow tungstic acid precipitate appears when concentrated hydrochloric acid begins to put into solution. Continue add concentrated hydrochloric acid until the pH value of solution is 2 and keep the pH unchanged for 30 minutes. Cool down the mixture.

(2) Acidize, diethyl ether extracts to prepare dodecatungstosilicic acid

Divert the solution to separating funnel, add diethyl ether (half of mixture volume). Add total 10ml concentrated hydrochloric acid for four times. After Stirring and extraction, lay it for a while to form three distinct layers. Upper layer is ether that mixes few heteropoly acid, middle layer is water solution of sodium chloride, hydrochloric acid and other substance, lower layer is oil heteropoly diethyl ether adduct. Separate the lower layer and put them into evaporating dish, add 4ml water, heat the mixture in the water-bath until crystal appears on the surface of solution, cool down to crystallize, suction filtration to acquire the production.

2. Thermal gravimetric curve (TG) and the differential thermal curve (DTA)

The TG analysisand differential thermal analysis are carried out on the thermogravimetric analyzer from room temperature to 650℃. Calculate the water content of sample to make sure the number of crystal water.

3. UV absorption spectrum

Prepare 5×10^{-5} mol/L dodecatungstosilicic acid into 1cm cuvette and take water as a blank, measure UV absorption spectrum on UV – 240 spectrophotometer, record the absorb curve from wavelength 200nm to 400nm.

4. Infrared spectroscopy

The samples was pressed to tablet with KBr, record the infrared spectroscopy from 400—4000cm^{-1} wavelength, and signal the major absorption peak.

Vocabulary

dodecatungstosilicic acid	十二钨硅酸	extraction	萃取
thermogravimetric synthesis	热重分析	vanadium	钒
niobium	铌	molybdenum	钼
tungsten	钨	isopoly acid	同多酸
heteropoly acid	杂多酸		

(编写：段丽颖，英文核审：夏丹丹)

实验二十六　三氯化六氨合钴（Ⅲ）的制备及组成的测定

【预习内容】

1. 钴（Ⅱ）、钴（Ⅲ）化合物的性质。
2. 沉淀滴定法。
3. 电导、电导率、摩尔电导率、稀度及 DDS – 11A 型电导率仪的使用。

4. 能全部解离的配合物的解离类型与摩尔电导率之间的实验规律。

【思考题】

1. 在[Co(NH$_3$)$_6$]Cl$_3$的制备过程中，NH$_4$Cl、活性碳和H$_2$O$_2$各起什么作用？影响产率的关键是什么？

2. 氨的测定原理是什么？氨测定装置中，漏斗下端插入小试管在NaOH液面下以及橡皮塞切口的原因是什么？

3. 测定钴含量时，样品液中加入10% NaOH溶液，加热产生棕黑色沉淀，这是什么化合物？用什么方法测定钴含量？

4. 氯的测定原理是什么？CrO_4^{2-}的浓度和溶液的酸度对分析结果有何影响？合适的条件是什么？

5. 何谓稀度？如配250ml稀度为128的[Co(NH$_3$)$_6$]Cl$_3$溶液，计算应准确称取该配合物的量。

6. 如何测定[Co(NH$_3$)$_6$]Cl$_3$的解离类型？

【实验原理】

根据有关电对的标准电极电势，在通常情况下，钴(Ⅱ)盐比钴(Ⅲ)盐稳定得多，而在形成配合物的情况下则相反，钴(Ⅲ)的配合物要比钴(Ⅱ)的配合物稳定得多。因此制备钴(Ⅲ)配合物时，常用钴(Ⅱ)化合物为原料，通过氧化反应来制备。如橙色的[Co(NH$_3$)$_6$]Cl$_3$的制备条件是以活性碳为催化剂，用H$_2$O$_2$氧化有NH$_3$和NH$_4$Cl存在的CoCl$_2$溶液。反应式为：

$$2CoCl_2 + 2NH_4Cl + 10NH_3 + H_2O_2 \xrightarrow{C} 2[Co(NH_3)_6]Cl_2 + 2H_2O$$

所得产品[Co(NH$_3$)$_6$]Cl$_3$为橙色单斜晶体，20℃时在水中的溶解度为0.26mol/L。

【仪器与试剂】

锥形瓶，布氏漏斗，抽滤瓶，蒸馏装置，滴定管，DDS-11A型电导率仪。

NH$_4$Cl(s)，CoCl$_2$·6H$_2$O(s)，KI(s)，活性碳，浓氨水，浓HCl，H$_2$O$_2$溶液(5%)，NaOH溶液(10%)，甲基红溶液(0.1%)，Na$_2$S$_2$O$_3$溶液(0.05mol/L)，淀粉溶液(0.2%)，AgNO$_3$溶液(0.1mol/L)，K$_2$CrO$_4$溶液(5%)。

【实验内容】

1. 三氯化六氨合钴(Ⅲ)的制备 在锥形瓶中将6g NH$_4$Cl(s)溶于10ml水中，加入9g研细的CoCl$_2$·6H$_2$O(s)和0.5g活性碳，摇动锥形瓶，使其混合均匀。冷却后加20ml浓氨水，进一步冷却到10℃以下，缓慢加入20ml 5% H$_2$O$_2$溶液，水浴加热至60℃左右，保持20分钟，并不断摇动锥形瓶，然后用冰浴冷却到0℃左右，用布氏漏斗抽滤。把沉淀溶于50ml沸水中（水中含3ml浓HCl），趁热抽滤。缓慢加入10ml浓HCl于滤液中，即有大量橘黄色晶体析出，用冰浴冷却后过滤。晶体用冷的2ml 2mol/L HCl洗涤，再用少许乙醇洗涤后吸干。将晶体在低于105℃条件下烘干。称量、计算产率。

2. 三氯化六氨合钴（Ⅲ）组成的测定

（1）氨的测定　准确称取 0.2g 左右的试样，放入 250ml 锥形瓶中，加 80ml 水溶解，然后加入 10ml 10% NaOH 溶液。在另一锥形瓶中准确加入 30～35ml 0.5mol/L HCl 标准溶液，放在冰浴中冷却。

按图 26-1 装配仪器，从漏斗加 3～5ml 10% NaOH 溶液于小试管中，漏斗柄下端插入液面 2～3cm。加热试液，开始可用大火，溶液近沸时改为小火，保持微沸状态，蒸馏 1 小时左右即可将溶液中的氨全部蒸出。蒸馏完毕，取出插入 HCl 溶液中的导管，用纯化水将导管内外冲洗（洗涤液全部流入氨吸收瓶中）。取出吸收瓶，加 2 滴 0.1% 甲基红溶液，用 0.5mol/L NaOH 标准溶液滴定过剩的 HCl，计算氨的含量。

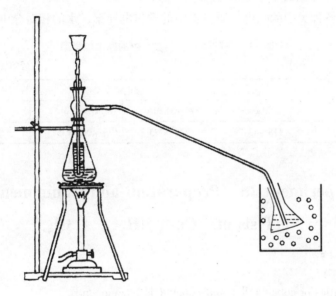

图 26-1　蒸氨装置

（2）钴的测定　准确称取 0.2g 左右（0.17～0.22g）试样两份分别置于 250ml 烧杯中，加水溶解。加入 10ml 10% NaOH 液。将烧杯放在水浴上加热。待氨全部赶走后（如何检查？）加入 1g KI（s）及 10ml 6mol/L HCl，于暗处放置 5 分钟左右。用 0.05mol/L 的标准 $Na_2S_2O_3$ 溶液滴定至浅黄色，加入 2ml 新配制 0.2% 淀粉溶液后，再滴至蓝色消失。计算钴的含量。

（3）氯的测定

① 配制 0.1mol/L $AgNO_3$ 标准溶液，计算滴定所需的试样量。

② 准确称取试样两份于锥形瓶中，分别加水 25ml，配成试液。

③ 加 1ml 5% K_2CrO_4 溶液为指示剂，用 0.1mol/L $AgNO_3$ 标准溶液滴定至出现淡红棕色不再消失为终点。

④ 由滴定数据，计算氯的含量。

由以上情况分析氨、钴、氯的结果，写出产品的实验式。

3. 三氯化六氨合钴（Ⅲ）解离类型的测定

（1）配制 250ml 稀度为 128 的试液，再用此溶液配制稀度分别为 256、512、1024

的试液 100ml。所谓稀度即溶液的稀释程度，为物质的量浓度的倒数，如稀度为 128，表示 128L 溶液中有 1mol 的溶质。用 DDS – 11A 型电导率仪测定溶液的电导率 γ。

(2) 按下列公式计算摩尔电导率，确定 [Co(NH$_3$)$_6$]Cl$_3$ 的解离类型，写出化学式。

$$\Lambda_m = \frac{\gamma \cdot 10^{-3}}{c}$$

【实验数据的记录及处理】

自行设计实验报告格式并进行数据处理。

附录

表 26 – 1 为各种类型的强电解质在水中的摩尔电导率，水的相对介电常数为 78.4。

表 26 – 1　各种强电解质在水中的摩尔电导率

(25℃，浓度 10^{-3} mol/L)

	强电解质类型			
	MA	M$_2$A 或 MA$_2$	M$_3$A 或 MA$_3$	M$_4$A 或 MA$_4$
Λ_m (S·m^2/mol)	118 ~ 131	235 ~ 273	408 ~ 435	500 ~ 560

Experiment 26　Preparation and Component Analysis of [Co(NH$_3$)$_6$]Cl$_3$

Preview

1. The properties of cobalt (Ⅱ) and cobalt (Ⅲ) compounds.

2. Precipitate titration method.

3. The definition of conduction, conductivity, molar conductivity and dilution; the usage of DDS-11A type conductivity meter.

4. The experimental regularity between dissociation type of completely dissociated complexes and molar conductivity.

Questions

1. What are the roles of NH$_4$Cl, H$_2$O$_2$ and active carbon in the process of preparation of [Co(NH$_3$)$_6$]Cl$_3$? What plays the key role in affecting yield?

2. What is the principle for determination of ammonia? In the device for determination of ammonia, why is there a small tube inserted under NaOH fluid level at the bottom of funnel? Why is a notched rubber plug used?

3. When 10% NaOH is added to the sample solution for determination of cobalt content, dark brown precipitate will appear after heating, what is this compound? What method is used to determine cobalt content?

4. What is the principle for determination of chlorine? What is the effect of the concentra-

tion of CrO_4^{2-} and the acidity of solution on the result of analysis? What is the appropriate condition?

5. What is dilution? What quantity of the complexes should be weighed to prepare 250ml $[Co(NH_3)_6]Cl_3$ solution which has the dilution of 128?

6. How to determine the dissociation type of $[Co(NH_3)_6]Cl_3$?

Principles

According to standard electrode potential, bivalent cobalt salt is more stable than trivalent cobalt salt, but in the case of their coordinates, most of cobalt(Ⅲ) complexes are more stable than cobalt(Ⅱ) complexes, so cobalt(Ⅲ) complexes are often prepared by oxidation reaction.

In this experiment, active carbon is used as catalyst, and hydrogen peroxide is used as oxidizer. Cobalt dichloride reacts with excessive ammonia and ammonium chloride to yield aqueous $[Co(NH_3)_6]Cl_3$. The main reaction is shown as follows:

$$2CoCl_2 + 2NH_4Cl + 10NH_3 + H_2O_2 \xrightarrow{C} 2[Co(NH_3)_6]Cl_2 + 2H_2O$$

$[Co(NH_3)_6]Cl_3$ is orange monoclinic crystal, its solubility in water is 0.26mol/L at 20℃.

Instruments and Chemicals

Erlenmeyer flask, Buchner funnel, suction flask, distillation apparatus, burette, DDS-11A type conductivity meter.

$NH_4Cl(s)$, $CoCl_2 \cdot 6H_2O(s)$, $KI(s)$, activated carbon, strong aqua ammonia, concentrated hydrochloric acid, H_2O_2 solution (5%), NaOH solution (10%), methyl red solution (0.1%), $Na_2S_2O_3$ solution (0.05mol/L), starch solution (0.2%), $AgNO_3$ solution (0.1mol/L), K_2CrO_4 solution (5%).

Procedures

1. Preparation of $[Co(NH_3)_6]Cl_3$

Add 6g NH_4Cl and 10ml purified water in an Erlenmeyer flask, then add 9g grinded $CoCl_2 \cdot 6H_2O$ and 0.5g active carbon, shake the Erlenmeyer flask to mix them evenly. Cool and add 20ml concentrated ammonia, cool to below 10℃, add 20ml 5% H_2O_2 slowly, heat by water bath to about 60℃ and keep the temperature for 20min, shaking the Erlenmeyer flask properly, then cool to about 0℃ by ice water, filter under reduced pressure. Dissolve the precipitate in 50ml boiling water which containing 3ml concentrated hydrochloric acid. Filter while it is hot, add 10ml concentrated hydrochloric acid slowly into filtrate, separate a lot of orange crystal out, filter after cooling by ice water. Wash the crystal with 2ml 2mol/L cool hydrochloric acid, then with a small amount of ethanol. Bake it below 105℃, weigh and calculate the yield.

2. Determination of components of $[Co(NH_3)_6]Cl_3$

(1) Determination of ammonia

Add 0.2g sample in 250ml Erlenmeyer flask, add 80ml purified water to dissolve the sample, then add 10ml 10% NaOH solution. Add 30—35ml 0.5mol/L HCl standard solution to other Erlenmeyer flask (accepter), immerge accepter in ice water.

Install the apparatuses according to Fig. 26-1. Add 3—5ml 10% NaOH in a small tube with a funnel, insert funnel handle under liquid level 2—3cm. Heat up sample solution to boiling, then reduce heating, to keep the solution simmering for about 1h, vaporing ammonia out completely. Wash the hydrochloric acid solution which perhaps adhered to inside and outside pipe into accepter with purified water, add 2 drops of methyl red indicator in the accepter, titrate the left hydrochloric acid by 0.5mol/L standard NaOH solution, calculate the content of ammonia.

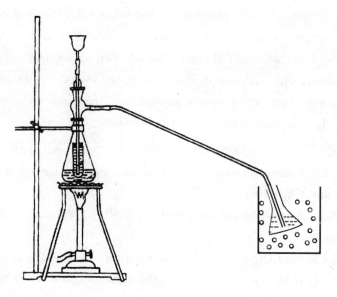

Fig. 26-1 Schematic of ammonia distillation

(2) Determination of cobalt

Weigh 0.2g(0.17—0.22g) sample in 250ml beaker, dissolve it with purified water, add 10ml 10% NaOH in beaker. Heat the beaker in water bath, after getting rid of ammonia (how to determine?), add 1g solid KI and 10ml 6mol/L HCl, lay aside for 5min in dark. Titrate it by 0.05mol/L standard $Na_2S_2O_3$ solution until solution becomes primrose yellow, add 2ml fresh 0.2% starch solution, titrate again until the blue disappears. Calculate the content of cobalt.

(3) Determination of chlorine

① Prepare 0.1mol/L $AgNO_3$ standard solution, calculate the quantity of sample required for titration.

② Weigh two samples in beakers, dissolve with 25ml purified water respectively.

③ Add 1ml 5% K_2CrO_4 solution as indicator, titrate it by 0.1mol/L $AgNO_3$ standard solution until the reddish brown no longer disappears.

④ Calculate the content of chlorine according to the titration data.

Determine the empirical formula of product according to the analysis result of ammonia, cobalt and chlorine.

3. Determination of the dissociation type of $[Co(NH_3)_6]Cl_3$

(1) Prepare 250ml $[Co(NH_3)_6]Cl_3$ solution which has the dilution of 128, prepare

100ml solution which has the dilution of 256, 512 and 1024 respectively. Dilution is the dilute strength of the solution, the reciprocal of molality. The dilution of 128 means that there is 1mol solute in 128L solution. Measure the electrical conductivity of solution γ using DDS-11A type conductivity meter.

(2) Calculate molar conductivity according to the formula:

$$\Lambda_m = \frac{\gamma \cdot 10^{-3}}{c}$$

Determine the dissociation type of $[Co(NH_3)_6]Cl_3$, write the chemical formula.

Data record and processing

Design the experiment report format and process data by yourself.

Appendix

Molar conductivity of strong electrolytes of various types in water (the relative dielectric constant in water is 78.4) is shown as table 26-1.

Table 26-1 Molar conductivity of strong electrolytes of various types

(25℃, Concentration 10^{-3} mol/L)

	Type of strong electrolyte			
	MA	M_2A or MA_2	M_3A or MA_3	M_4A or MA_4
$\Lambda_m(S \cdot m^2/mol)$	118—131	235—273	408—435	500—560

Vocabulary

conductivity	电导率	dilution	稀度
dissociation type	解离类型	precipitate titration	沉淀滴定法
complexes	配合物	monoclinic crystal	单斜晶体
catalyst	催化剂	empirical formula	实验式
strong electrolytes	强电解质	relative dielectric constant	相对介电常数

（编写：刘晶莹，英文核审：肖琰）

实验二十七　植物中某些元素的分离与鉴定

【预习内容】

1. Ca^{2+}、Mg^{2+}、Fe^{3+}、I^-和PO_4^{3-}的鉴定反应。
2. 植物样品的灰化及浸溶。

【思考题】

1. 如何用控制酸度的方法分离Ca^{2+}、Mg^{2+}和Fe^{3+}？
2. 试液中Ca^{2+}、Mg^{2+}和Fe^{3+}共存，请设计一份合理、简洁的分离鉴定方案。

【实验原理】

植物是有机体，主要由C、H、O、N等元素组成，此外，还含有P、I和某些金属元素如 Ca、Mg、Fe 等。把植物烧成灰烬，然后用酸浸溶，即可从中分离和鉴定某些元素。本实验只要求分离和检出植物中 Ca、Mg、Fe 三种金属元素和 P、I 两种非金属元素。

【仪器与试剂】

电子天平，煤气灯，蒸发皿，研钵，烧杯（100ml），药匙，广范 pH 试纸。

酸：HCl（2mol/L），HCl（稀），H_2SO_4（2mol/L），HNO_3（浓），HAc。

碱：$NH_3 \cdot H_2O$（浓），NaOH（2mol/L）。

盐：$(NH_4)_2C_2O_4$（0.5mol/L），NH_4SCN，$(NH_4)_6Mo_7O_{24}$，KNO_2。

其他：镁试剂，氯水（TS），四氯化碳，松枝，柏枝，茶叶，海带。

【操作步骤】

1. 植物材料的准备

（1）松枝、柏枝的枯枝或干叶，由同学自己采集。

（2）茶叶、海带，由实验室统一准备。

2. 从茶叶、松枝、柏枝中鉴定 Ca、Mg、Fe

（1）取约 10g（青叶为 20g）已干燥的植物材料，放入蒸发皿中，用煤气灯加热进行灰化（在通风橱中进行）。

（2）充分灰化后，移入研钵磨细，将 15ml 浓度为 2mol/L 的盐酸加入研钵的灰中搅拌。

（3）过滤盐酸溶液，得滤液（Ⅰ），滤液（Ⅰ）中金属离子按 Ca、Mg、Fe 顺序检出。

（4）Ca^{2+} 及 Mg^{2+} 的检出　往滤液（Ⅰ）中加入浓氨水，调 pH≤8，待有沉淀（Ⅰ）生成，过滤，得滤液（Ⅱ）。

将滤液（Ⅱ）分成两份，在其中一份溶液中，加入几滴 0.5mol/L 草酸铵试液，若有白色沉淀生成，分离，沉淀不溶于醋酸，但可溶于稀盐酸，证明有 Ca^{2+} 存在。

$$Ca^{2+} + C_2O_4^{2-} = CaC_2O_4 \downarrow （白色）$$

$$CaC_2O_4 + 2H^+ = Ca^{2+} + H_2C_2O_4$$

在另一份溶液中加入镁试剂（0.001% 对硝基苯偶氮 α-萘酚的 NaOH 溶液 2mol/L），若出现蓝色沉淀，或溶液由紫色变为蓝色，证明有 Mg^{2+} 存在。

$$1/2Mg^{2+} + \text{H}_2\text{O}\underset{\text{OH}}{\bigcirc}-N=N-\bigcirc-NO_2 \xrightarrow{NaOH} \text{HO}\underset{\text{O-Mg/2}}{\bigcirc}-N=N-\bigcirc-NO_2 \downarrow$$

$$(\text{blue})$$

（5）Fe^{3+} 的检出：将滤出的沉淀（Ⅰ）移入小烧杯中，加入 10ml 浓度为 2mol/L 的氢氧化钠溶液搅拌，过滤，得沉淀（Ⅱ）。

取沉淀（Ⅱ）加约 10ml 浓度为 2mol/L 的硫酸溶解，分为两等份。在其中一份溶液中加入几滴 0.1% 的亚铁氰化钾溶液，若有蓝色沉淀生成，表示有 Fe^{3+} 存在；在另一份溶液中加入 1 滴硫氰酸铵溶液，若显血红色，也表示有 Fe^{3+} 存在。

$$K^+ + Fe^{3+} + [Fe(CN)_6]^{4-} = KFe[Fe(CN)_6]\downarrow \text{（蓝色）}$$
$$Fe^{3+} + nSCN^- = Fe(SCN)_n^{3-n}\downarrow \text{（血红色）} \quad (n = 1 \sim 6)$$

3. 植物中磷的鉴定

（1）植物灰的制备（同前）。

（2）取一药匙灰（约 0.5g）于 100ml 烧杯中，加 2ml 浓 HNO_3 溶液，再加入 30ml 纯化水，过滤得透明溶液。

（3）在滤液中加入 1ml H_2SO_4 和钼酸铵（$H_8MoN_2O_4$）的混合液（在使用之前，将 4 体积 50% 的硫酸和 1 体积 10% 的钼酸铵混合而成），若生成黄色沉淀，加 $NH_3 \cdot H_2O$ 及 NaOH 沉淀溶解，则表示 PO_4^{3-} 存在。

$$HPO_4^{2-} + 3NH_4^+ + 12MoO_4^{2-} + 23H^+ = (NH_4)_3[P(Mo_3O_{10})_4] \cdot 6H_2O\downarrow \text{（黄色）} + 6H_2O$$

4. 海带中碘元素的鉴定

（1）取 10g 左右的海带进行灰化（在通风橱中进行）。

（2）取一药匙灰（约 0.5g）于 100ml 烧杯中，加入 10ml 5% 的醋酸溶液，稍加热解，过滤。

（3）在滤液中，滴加氯水或亚硝酸钾固体，充分摇匀。

（4）在此溶液中加入等体积的四氯化碳，充分振荡后静置，若四氯化碳层中有玫瑰红色出现，表示有 I^- 存在。

$$2I^- + 2NO_2^- + 4H^+ = I_2 + 2NO + 2H_2O$$
$$2I^- + Cl_2 = I_2 + 2Cl^-$$

实验数据的记录与说明：

（1）本实验可根据自选植物标本进行。但应注意各种植物中，以上几种元素的含量不尽相同。将实验结果填入下列表格中。

元素名称＼植物名称	松枝	柏枝	茶叶	海带
钙				
镁				
铁				
磷				
碘				

检出该元素，填写为"＋"，在被测液相同量的情况下，视检出的多少，可填写为"＋，＋＋，＋＋＋"，没有检出时填写"－"。

（2）三种金属离子的氢氧化物完全沉淀的 pH 范围为：

$Ca(OH)_2$ pH＞13；$Mg(OH)_2$ pH＞11；$Fe(OH)_3$ pH≥4.1

开始分离时，加氨水控制 pH≥8，此时钙、镁氢氧化物不沉淀，而铁氢氧化物沉淀。

Experiment 27 Isolation and Identification of Inorganic Elements in Plants

Preview

1. Identification of Ca^{2+}, Mg^{2+}, Fe^{3+}, I^- and PO_4^{3-}.
2. Dry ashing of plant samples and extraction of minerals.

Questions

1. How to separate cations of Ca^{2+}, Mg^{2+} and Fe^{3+} by pH adjustment?
2. Please design a feasible method for isolation and identification of Ca^{2+}, Mg^{2+} and Fe^{3+} in a solution.

Principles

Plants are organic, mainly made of C, H, O and N. Besides these macro-elements, plants also contain P, I and metal elements such as Ca, Mg and Fe. Some of the inorganic elements can be extracted and identified after being burned to ash. This experiment requires you to isolate and identify of the three metal elements of Ca, Mg and Fe and two non-metal elements, P and I.

Instruments and Chemicals

electronic balance, Bunsen burner, evaporating dish, mortar, beakers (100ml), dispensing spoon, universal pH indicator.

acid: HCl (2mol/L), HCl (dilute), H_2SO_4 (2mol/L), HNO_3 (concentrated), HAc.

alkali: $NH_3 \cdot H_2O$ (concentrated), NaOH (2mol/L).

salt: $(NH_4)_2C_2O_4$ (0.5mol/L), NH_4SCN, $(NH_4)_6Mo_7O_{24}$, KNO_2.

others: magneson ii, CCl_4, branches of pine and cypress, tea, kelp.

Procedure

1. Preparation of plant materials

(1) Collect litter of pine and cypress by yourself.

(2) Tea and kelp will be provided by the inorganic laboratory.

2. Identification of Ca, Mg and Fe from tea, pine and cypress samples

(1) Add about 10g of dried plant material (fresh leaf: 20g) in to an evaporating dish, ashing with a Bunsen burner. This step should be performed in a fume hood.

(2) Transfer the completely ashed residue to a mortar and grind the material. Add 15ml of 2mol/L hydrochloric acid into the finely ground ash and stir the mixture.

(3) Filter the hydrochloric acid solution with filter paper. Identify the metal ions of Ca, Mg and Fe from the filtrate (I) successively.

(4) Test for Ca^{2+} and Mg^{2+}

Add concentrated ammonia to the filtrate (I) to adjust pH value not more than 8. The precipitate will form. Filter the precipitate (Ⅰ) to obtain filtrate (Ⅱ).

Divide the filtrate (Ⅱ) into two aliquots. Add a few drops of 0.5mol/L ammonium oxalate to the first aliquot. The formation of a white precipitate, which is soluble in hydrochloric acid but insoluble in acetic acid, is indicative of the presence of cation Ca^{2+}.

$$Ca^{2+} + C_2O_4^{2-} = CaC_2O_4 \downarrow (\text{white})$$
$$CaC_2O_4 + 2H^+ = Ca^{2+} + H_2C_2O_4$$

Add magneson ii [4-(4-Nitrophenylazo)-1-naphthol] (0.001% in 2mol/L sodium hydroxide) to the other aliquot. The formation of blue precipitate or the change in the color of the solution from purple to blue is indicative of the presence of cation Mg^{2+}.

$$1/2Mg^{2+} + H_2O\text{-}\underset{OH}{\bigcirc}\text{-}N\text{=}N\text{-}\bigcirc\text{-}NO_2 \xrightarrow{NaOH} HO\text{-}\bigcirc\text{-}N\text{=}N\text{-}\bigcirc\text{-}NO_2 \downarrow$$
$$\underset{(\text{blue})}{O\text{-}Mg/2}$$

(5) Test for Fe^{3+}

Transfer the precipitate (Ⅰ) into a 50ml beaker. Add 10ml of 2mol/L sodium hydroxide solution to the beaker, followed by stirring and filtration, affording precipitate (Ⅱ).

Dissolve precipitate (Ⅱ) in 10ml of 2mol/L sulfuric acid and divide the solution into two aliquots. Add a few drops of 0.1% potassium ferrocyanide to one aliquot. Formation of blue precipitate is indicative of presence of Fe^{3+}. Add one drop of ammonium thiocyanate solution to the other aliquot. If the test solution contains Fe^{3+}, the resulting mixture will turn into blood red color.

$$K^+ + Fe^{3+} + [Fe(CN)_6]^{4-} = KFe[Fe(CN)_6] \downarrow (\text{blue})$$
$$Fe^{3+} + nSCN^- = Fe(SCN)_n^{3-n} \downarrow (\text{blood red}) \quad (n = 1\text{—}6)$$

3. Identification of P in plant material

(1) Ashing of the plant material as described in section 2.

(2) Successively add 2ml of concentrated nitric acid and 30ml of purified water into a 100ml beaker containing 0.5g of plant ash. Filter the solution to give a clear filtrate.

(3) Add 1ml of the mixed solution of sulfuric acid and ammonium molybdate which can be prepared right before use by mixing four volumes of 50% sulfuric acid with one volume of 10% ammonium molybdate solution. Formation of a yellow precipitate that is soluble in $NH_3 \cdot H_2O$ or NaOH solutions is indicative of the presence of PO_4^{3-}.

$$HPO_4^{2-} + 3NH_4^+ + 12MoO_4^{2-} + 23H^+ = (NH_4)_3[P(Mo_3O_{10})_4] \cdot 6H_2O \downarrow (\text{yellow}) + 6H_2O$$

4. Identification of I in kelp

(1) Prepare the ash from 10g of kelp in a fume hood.

(2) Add 10ml of 5% acetic acid solution to a 100ml beaker containing 0.5g of kelp ash. After heating the solution somewhat, filter the solution through filter paper.

(3) Drop some chlorine solution or solid potassium nitrite into the filtrate and shake the solution well.

(4) Add equivalent volume of carbon tetrachloride. Shake the mixture well, allow it to stand and produce two separated solvent layers. A rose red lower layer indicates the presence of I^- in the test sample.

$$2I^- + 2NO_2^- + 4H^+ = I_2 + 2NO + 2H_2O$$
$$2I^- + Cl_2 = I_2 + 2Cl^-$$

Data record and processing

(1) You can choose plant material by yourself. Please note that the contents of the elements we discussed above may vary significantly in different plant materials. Fill the following form with your experiment results.

Element \ Plant material	Branches of Pine	Branches of cypress	Tea	Kelp
Calcium				
Magnesium				
Ferrum				
Phosphorus				
Iodine				

Positive reaction: +; negative reaction: −; numbers of + means the relative content of the test element.

(2) The pH ranges for completely precipitate the hydroxides of the three metal cations of Ca^{2+}, Mg^{2+} and Fe^{3+}:

$$Ca(OH)_2 \quad pH>13; \quad Mg(OH)_2 \quad pH>11; \quad Fe(OH)_3 \quad pH \geq 4.1$$

Ammonia can be used to make the $pH \geq 8$, therefore allow Fe^{3+} to precipitate whereas Ca^{2+}, Mg^{2+} soluble.

Vocabulary

buret	滴定管	ammonium thiocyanate	硫氰酸铵
volumetric pipet	移液管	sulfuric acid	硫酸
volumetric flask	容量瓶	ammonium molybdate	钼酸铵
chlorine solution	氯水	volumetric glassware	滴定管
evaporating dish	蒸发皿	potassium nitrite	亚硝酸钾
inorganic elements	无机元素	carbon tetrachloride	四氯化碳
calcium	钙	magnesium	镁
magneson ii	镁试剂	phosphorus	磷
ammonium oxalate	草酸铵	iodine	碘
acetic acid	醋酸	potassium ferrocyanid	亚铁氰化钾

(编写：凌俊红，英文核审：肖琰)

附　录

一、常用的酸碱指示剂

指示剂	变色范围	颜色		配制方法
		酸式色	碱式色	
百里酚蓝	1.2～2.8	红	黄	0.1g 指示剂溶于 100ml 20% 乙醇溶液
甲基黄	2.9～4.0	红	黄	0.1g 指示剂溶于 100ml 90% 乙醇溶液
甲基橙	3.1～4.4	红	黄	0.1% 的水溶液
溴酚蓝	3.0～4.6	黄	蓝紫	0.1g 指示剂溶于 100ml 20% 乙醇溶液或其钠盐水溶液
甲基红	4.4～6.2	红	黄	0.1g 或 0.2g 指示剂溶于 100ml 60% 乙醇溶液
溴麝香草酚蓝	6.2～7.6	黄	蓝	0.1g 指示剂溶于 100ml 20% 乙醇溶液或其钠盐水溶液
中性红	6.8～8.0	红	黄	0.1g 指示剂溶于 100ml 60% 乙醇溶液
酚红	6.4～8.0	黄	红	0.1g 指示剂溶于 100ml 60% 乙醇溶液或其钠盐水溶液
酚酞	8.0～10.0	无	红	0.1g 指示剂溶于 100ml 60% 乙醇溶液或 1g 指示剂溶于 100ml 90% 乙醇溶液
百里酚酞	9.4～10.6	无	蓝	0.1g 指示剂溶于 100ml 90% 乙醇溶液
刚果红	3.0～5.2	蓝紫	红	0.1% 的水溶液
溴酚红	5.0～6.8	黄	红	0.1g 或 0.04g 指示剂溶于 100ml 20% 乙醇溶液

二、常用的酸碱密度和浓度

酸或碱	分子式	密度（g/cm³）	质量分数	浓度（mol/L）
冰醋酸	CH_3COOH	1.05	0.995	17
浓盐酸	HCl	1.18	0.36	12
浓硝酸	HNO_3	1.42	0.72	16
浓硫酸	H_2SO_4	1.84	0.96	18
磷酸	H_3PO_4	1.69	0.85	15
浓氨水	$NH_3 \cdot H_2O$	0.90	0.28～0.30	15

三、酸度计简介

酸度计又称 pH 计，属于一种电化学测量仪，主要通过测量电势差的方法测定溶液的 pH。实验室中常见的酸度计型号有 pHS-25、pHS-2、pHS-3 等，虽然它们结构各异，但基本原理相同。

（一）基本原理

将一对电极插入被测溶液中，不同的 pH 值溶液，会在电极间产生不同的电动势值；酸度计内部的转换器可将电动势值转换成 pH 值，并直接显示在仪器的读数屏上。

上述的对电极通常采用玻璃电极和饱和甘汞电极（附图 3-1）。玻璃电极为测量电极，其电极电势随被测溶液的 pH 值不同而变化；饱和甘汞电极为参比电极，其电极电势为恒定值，与被测溶液的 pH 值无关，25℃时，饱和甘汞电极电势为 0.2415V。

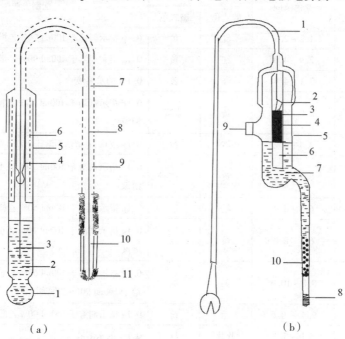

附图 3-1 玻璃电极和饱和甘汞电极

（a）玻璃电极　　　　　　　　　　（b）饱和甘汞电极

1. 玻璃薄膜；2. 缓冲溶液；3. Ag-AgCl 电极；　1. 电极引线；2. 玻璃管；3. 汞；4. 甘汞糊；
4. 电极导线；5. 玻璃管；6. 静电隔离层；　　5. 玻璃外套；6. 石棉或纸浆；7. 饱和 KCl 溶液；
7. 电极导线；8. 塑料绝缘线；9. 金属隔离罩；　8. 素烧瓷；9. 加液孔；10. KCl 晶体
10. 塑料绝缘线；11. 电极接头

玻璃电极头部的薄玻璃膜对氢离子十分敏感，当其浸入被测溶液后，被测溶液中的氢离子与玻璃球泡表面水化层进行离子交换而产生电极电势，随被测溶液 pH 值不同，玻璃电极的电势值随之改变。

在25℃时，玻璃电极的电势为：

$$\varphi（玻璃）= \varphi^{\ominus}（玻璃）+ 0.0592 \lg [H^+] = \varphi^{\ominus}（玻璃）- 0.0592 \text{pH}$$

饱和甘汞电极符号为：

$$Hg(l) | Hg_2Cl_2(s) | KCl（饱和）$$

电极反应：　　　　　　$Hg_2Cl_2 + 2e = 2Hg + 2Cl^-$

将玻璃电极和饱和甘汞电极浸入被测溶液中组成原电池，并连接精密的电位计，

即可测定原电池的电动势,电动势 E 与被测溶液的 pH 值关系为:

$E = \varphi(正) - \varphi(负) = \varphi(饱和甘汞) - \varphi(玻璃) = 0.2415 - \varphi^{\ominus}(玻璃) + 0.0592\,\mathrm{pH}$

在25℃时,$\varphi^{\ominus}$(玻璃)可用已知 pH 的缓冲溶液(附表3-1)代替待测溶液测定电动势,进而求得。

为了保护电极和方便操作,现在的酸度计都配有 pH 复合电极(附图3-2),即将玻璃电极和饱和甘汞电极复合于一体。pHS-25型酸度计见附图3-3。

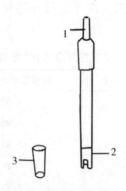

附图3-2 pH 复合电极
1. 电路插;2. 复合电极;3. 电极保护套

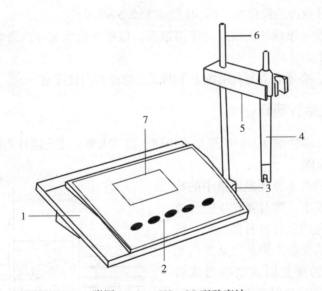

附图3-3 pHS-25型酸度计
1. 机箱;2. 按键;3. 电极梗固定座;4. 电极;5. 电极夹;6. 电极梗;7. 显示屏

(二)使用方法

(1)将复合电极固定在电极架上,取下复合电极上端的橡皮帽和下端的保护套。

(2)打开电源开关,仪器进入 pH 的测量状态。

(3)按"温度"键,使仪器进入温度调节状态,调节温度数值,使温度显示值与实际溶液温度一致,按"确认"键。

（4）将干净电极插入 pH = 6.86 的磷酸盐的标准缓冲溶液中，按"标定"键，显示 mV 值，按两次"确认"键，仪器进入"斜率"的标定状态。

（5）将干净电极插入到与待测溶液的液性一致的标准缓冲溶液中，如酸性待测液用 pH = 4.00（如碱性的用 pH = 9.18），显示 mV 值，按两次"确认"键，仪器进入 pH 的测量"笑脸"状态。

（6）将干净的电极插入待测溶液中，即可读取 pH。

（7）测量完毕，关闭电源开关，洗净电极并将复合电极上下两端的橡皮帽和橡皮套套上。

附表 3 – 1　标准缓冲溶液的 pH 值与温度对照表

温度/℃	苯二甲酸盐标准缓冲液	磷酸盐标准缓冲液	硼砂标准缓冲液
10	4.00	6.92	9.33
15	4.00	6.90	9.28
20	4.00	6.88	9.23
25	4.01	6.86	9.18
30	4.02	6.85	9.14

（三）注意事项

（1）仪器校正所用的缓冲溶液必须用 pH 值的基准试剂配制；配制标准缓冲溶液时，应用新沸过并放冷的纯化水，其 pH 值应为 5.5 ~ 7.0。

（2）每次更换标准缓冲溶液或测定溶液前，应用纯化水充分洗涤电极，并用滤纸将电极上的水吸干。

（3）测量时，必须取下复合电极上下两端的橡皮帽和橡皮套。

四、分光光度计简介

分光光度计有多种型号，但原理基本相同，以 721 型分光光度计为例，进行介绍。

（一）基本原理

分光光度计的基本原理是溶液中的物质在光的照射激发下，能对光产生吸收效应，由于物质对光的吸收具有选择性，所以物质都有特征的吸收光谱。当某单色光通过溶液时，光的能量因被吸收而减弱（附图 4 – 1），光能量减弱的程度、溶液的吸光度及吸光物质的浓度之间的关系符合朗伯 – 比尔定律：

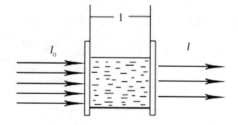

附图 4 – 1　单色光通过溶液

$$A = \lg \frac{I_0}{I} = \varepsilon l c$$

式中，A 为吸光度；I_0 为入射光强度；I 为透射光强度；ε 为摩尔吸光系数，在给定单色光、溶剂及温度等条件时，其为物质的特征常数，表示物质对某特定波长光的吸收能力；c 为吸光物质的浓度（mol/L）；l 为液层厚度（cm）。

当入射光强度、吸光系数及液层厚度一定时，溶液的吸光度只随溶液浓度而变化。

（二）使用方法

721 型分光光度计仪器外型如附图 4-2 所示。

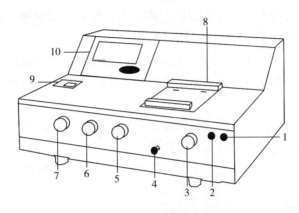

附图 4-2　721 型分光光度计

1. 电源指示灯；2. 电源开关；3. 灵敏度选择旋钮；4. 比色皿座定位拉杆；
5. 透光率 100% 电位器旋钮；6. 透光率 0 电位器旋钮；7. 波长调节器旋钮；
8. 比色皿暗箱；9. 波长示窗；10. 光密度示窗

1. 用分光光度计测定物质的最大吸收波长

（1）在仪器通电前，先调节调透光率 0 电位器旋钮 6，使光密度表的指针位于"0"刻线。

（2）接通电源，打开电源开关旋钮 2 和比色皿箱盖 8，预热约 20 分钟。

（3）调节波长调节器旋钮 7，选择某一波长，调节灵敏度选择旋钮 3 选择合适灵敏档；使装有参比溶液的比色皿处于空白校正位置，调透光率 0 电位器旋钮 6，使光密度表指示"0"；盖上比色皿暗箱盖，推进比色皿座定位拉杆 4，使参比溶液处于光路位置，调透光率 100% 电位器旋钮 5，使表针指向透光率 100% 附近。

（4）用纯化水、被测溶液分别润洗另一只比色皿后，装入被测溶液，放进比色皿架。按步骤（3）连续几次调整透光率"0"和"100%"，直至稳定。注意调"0"时打开暗箱盖，调"100%"时关闭暗箱盖。

（5）拉动比色皿架定位拉杆，使被测溶液处于光路中，在 400~800nm 范围内，每隔 10nm 测一次吸光度，但在最大吸收峰附近应每隔 5nm 测量一次。注意每选一个入射光波长时，需要重新调节"0"透光率及 100% 透光率，记录波长 λ 及吸光度 A。

（6）测量完毕，关闭电源，将比色皿洗净，用擦镜纸吸干，收入比色皿盒中。

（7）绘制 A-λ 曲线，找出被测液的最大吸收波长。

2. 用分光光度计测定被测溶液的浓度

（1）选择最大吸收波长为测定波长。

（2）仪器完成校正后，打开比色皿暗箱盖，除装有参比溶液的比色皿外，其余的比色皿可装入相应的待测溶液，放入比色皿架中，关闭比色皿暗箱盖；拉动比色皿架定位拉杆，使其处于光路中，测定它们的吸光度值。

(3) 测试完毕后，关闭电源，将比色皿洗净，用擦镜纸吸干，收入比色皿盒中。

(4) 根据标准溶液的 $A-c$ 曲线，找出被测溶液的对应浓度。

(三) 注意事项

(1) 当波长改变较大时，需要待稳定片刻后，再重新调整透光率和透光率，因为钨灯在急剧改变亮度后需要时间达到热平衡。

(2) 调节透光率于100%处，目的是提高消光读数，以适应高浓度溶液的测定。

(3) 选用合适光程长度的比色皿，使光密度读数处于0.1~0.65，目的是获得较高的准确度。

(4) 比色皿用过后要及时清洗，倒置晾干后存放在比色皿盒内。

(5) 灵敏度选择原则是：在保证空白溶液能调到透光率100%的前提下，尽可能采用较低档（一般放在"1"档），然后逐步增加，以获得较高的稳定性。

五、电导率仪

DDS—11A型电导率仪是目前最常用的电导率测量仪，广泛用于科研、生产、教学及环境保护等许多领域，除了可测量各种液体介质电导率外，还可以测量高纯水电导率。

(一) 基本原理

导体导电能力的大小，常用电阻或电导表示：

$$G = \frac{1}{R} \tag{1}$$

式 (1) 中，R 为电阻 (Ω)；G 为电导 (S)。

同金属导体一样，电解质溶液的电阻也符合欧姆定律。温度一定时，两极间溶液的电阻与两极间的距离 L 呈正比，与电极面积 A 呈反比：

$$R = \rho \frac{L}{A} \tag{2}$$

式 (2) 中，ρ 为电阻率，它的倒数称为电导率，以 γ (S/m) 表示。

将式 (2) 代入式 (1)，得：

$$\gamma = G \frac{L}{A} \tag{3}$$

电导率 γ 表示放在相距1m、面积为 $1m^2$ 的两个电极之间溶液的电导。

在电导池中，电极距离和面积是一定的，所以对某一电极来说，$\frac{L}{A}$ 是常数，称为电极常数或电导池常数，用 K 表示，则式 (3) 可表示为：

$$\gamma = K \cdot G \tag{4}$$

对于某一电极，电极常数 K 值是确定的，因此测定溶液的电导率 γ 可以得出溶液电导的大小。而溶液的导电能力的大小正比于溶液中电解质的含量，所以通过对电解质水溶液电导率的测量，可以计算溶液中电解质的含量。

(二)使用方法

DDS—11A 型电导率仪的面板示意图如附图 5-1。

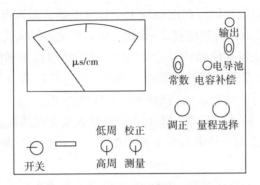

附图 5-1 DDS—11A 型电导率仪的外形图

其测量范围为 $0 \sim 10^5 \mu s/cm$，分 12 个量程，不同的量程配用不同的电极，不同量程的配用电极见附表 5-1。

附表 5-1 测量范围与配用电极

量程	电导率（μs/cm）	使用频率	配用电极
1	$0 \sim 0.1$	低周	DJS—1 型光亮电极
2	$0 \sim 0.3$		DJS—1 型光亮电极
3	$0 \sim 1$		DJS—1 型光亮电极
4	$0 \sim 3$		DJS—1 型光亮电极
5	$0 \sim 10$		DJS—1 型光亮电极
6	$0 \sim 30$		DJS—1 型铂黑电极
7	$0 \sim 10^2$		DJS—1 型铂黑电极
8	$0 \sim 3 \times 10^2$		DJS—1 型铂黑电极
9	$0 \sim 10^3$	高周	DJS—1 型铂黑电极
10	$0 \sim 3 \times 10^3$		DJS—1 型铂黑电极
11	$0 \sim 10^4$		DJS—1 型铂黑电极
12	$0 \sim 10^5$		DJS—10 型铂黑电极

具体使用步骤如下：

（1）电源开启前，观察表头指针是否指零。如不指零，调整表头上的调零螺丝，使指针指零。

（2）将校正、测量开关拨在"校正"位置。

（3）打开电源开关，预热 5~10 分钟，待指针稳定后调节调正器，使指针指示满刻度。

（4）根据待测液体电导率的大小，选择低周或高周。将开关拨向选定位置。

（5）将量程选择开关拨到所需要的测量范围档。如预先不知道待测液体的电导率范围，应先选择最大档，然后逐档下降至适宜档位，以防止指针被打弯。

（6）根据待测液体电导率的大小，选择配用电极。同时将电极常数调节器调到该电极标示的常数位置。例如：配用电极的标示常数为 0.95，则将常数调节器调至 0.95。

（7）用电极夹，将电极固定在电极杆上。将电极插头插入电极插口，旋紧插口上的固定螺丝。用少量待测液冲洗电极 2~3 次，然后将电极插入待测液中。

（8）再次调节调正器，使指针指示满刻度。然后将校正、测量开关拨至测量位置，读取指针示数，再乘上量程选择开关所指的倍率，即为待测溶液的实际电导率。例如：量程选择是 $0 \sim 10^2$，指针示数 0.9，则溶液的电导率为 90。

（9）选择量程 1、3、5、7、9、11 时，应读取表头上面的数值（0~1.0）；选择量程 2、4、6、8、10、12 时，读取表头下面的数值（0~3.0），即红点对红线，黑点对黑线。

（10）当用 $0 \sim 0.1 \mu s/cm$ 或 $0 \sim 0.3 \mu s/cm$ 测量高纯水时，先把电极引线插入电极插口，在电极未浸入溶液前，调节电容补偿调节器，使指针指示为最小值（由于电极之间存在漏电阻，致使调节电容补偿调节器时，指针不能达到零点），然后开始测量。

（11）测量完毕后，关闭电源，用纯化水冲洗电极后，放回电极盒中。

（三）注意事项

（1）电极使用前应在纯化水中浸泡，但不要弄湿电极引线，否则测量结果不准。

（2）为保证读数准确，应尽量使指针指示近于满刻度。

（3）当测量电导率大于 $10^4 \mu s/cm$ 时，应选用 DJS—10 型铂黑电极配用。注意此时电极常数调节器调节到该电极常数 1/10 的数值上。例如，若电极常数为 9.8，则应使调节器指在 0.98 处。将指针的读数乘以 10，即为被测液的电导率。

（4）高纯水被盛容器后要尽快测量，否则空气中的 CO_2 溶于水而解离出 H^+ 和 HCO_3^-，使电导率增大。

（编写：刘晶莹）

参 考 文 献

[1] 国家药典委员会. 中华人民共和国药典（第二部）[M]. 北京：中国医药科技出版社，2010.
[2] 国家药典委员会. 中华人民共和国药典（第二部）（英文版）[M]. 北京：中国医药科技出版社，2010.
[3] 伍晓春，姚淑心. 无机化学实验（英汉双语）[M]. 北京：科学出版社，2010.
[4] 刘静. 基础化学实验（英汉双语）[M]. 南京：东南大学出版社，2010.
[5] 曹凤岐. 无机化学实验与指导（英汉双语）[M]. 北京：中国医药科技出版社，2006.
[6] 张树彪，那立艳，华瑞年. 双语物理化学实验[M]. 北京：化学工业出版社，2009.
[7] 严拯宇. 分析化学实验与指导（英汉双语）[M]. 北京：中国医药科技出版社，2005.
[8] 古凤才，肖衍繁. 基础化学实验教程[M]. 北京：科学出版社，2003.
[9] 钟国清. 无机及分析化学实验[M]. 北京：科学出版社，2012.
[10] 包新华，邢彦军，李向清. 无机化学实验[M]. 北京：科学出版社，2013.
[11] 郑春生，杨南，李梅，等. 基础化学实验（无机及化学分析实验部分）[M]. 天津：南开大学出版社，2001.
[12] 王秋长，赵鸿喜，张守民，等. 基础化学实验[M]. 北京：科学出版社，2003.
[13] 北京师范大学无机化学教研室. 无机化学实验（第三版）[M]. 北京：高等教育出版社，2001.
[14] 郑春生，杨南，李梅. 基础化学实验[M]. 天津：南开大学出版社，2001.
[15] 崔学桂，张晓丽. 基础化学实验[M]. 北京：化学工业出版社，2003.
[16] 武汉大学化学与分子科学实验中心. 无机化学实验[M]. 武昌：武汉大学出版社，2002.
[17] 南京大学《无机及分析化学实验》编写组. 无机及分析化学实验（第三版）[M]. 北京：高等教育出版社，1998.